Attacks of Terror

Attacks of Terror

Surviving the Unthinkable

J. Brett Earnest

iUniverse, Inc.
New York Lincoln Shanghai

Attacks of Terror
Surviving the Unthinkable

iUniverse, Inc.

For information address:
iUniverse, Inc.
2021 Pine Lake Road, Suite 100
Lincoln, NE 68512
www.iuniverse.com

ISBN: 0-595-28304-7 (pbk)
ISBN: 0-595-74912-7 (cloth)

Printed in the United States of America

For Magdalena, whose love, devotion, and support make all things possible.

Contents

Acknowledgements

I want to thank Magdalena Hladka for sticking with me through the writing of this book. Without complaint and without reservation she has supported my weeks of research, my late nights, my early mornings, and my dedication to bringing this concise and current book to fruition. I also would like to thank her for using her hard-earned talents as a graphic designer to design the cover of this book.

I thank my family for their useful suggestions throughout this text. My mother, Phyllis, has been a guide for my use of language not only in this book, but throughout my life as well. My sister, Lynn, and my brothers, Chris and Craig, have provided timely and valuable input throughout these pages and over the years. Chris especially has offered numerous improvements for which the reader may be thankful. And to my late father, Joseph, whose view of the world, and the position of the United States in it, helped shape my belief in what it is to be an American.

Introduction

I wish there was no need for this book to be written. The topics I cover in this book are not easy to address. The task is made even less attractive when I know that sooner or later one of the deadly scenarios in the following pages will come to pass. On the other hand, it is a fact that terrorists will strike again in our country, which makes this book all the more critical. But how do you go about preparing someone for the unthinkable?

I searched the bookstands, the libraries, and the Internet in an attempt to find a straightforward, to-the-point book that would communicate exactly what options are available to a person swept up in an act of terror. I found several promising sources that, in the end, landed short of my expectations. More often than not, it seemed the sources would approach a topic head-on and then turn just in time to avoid confronting the ugly issues that inevitably attend the real world of terrorist attacks. Nevertheless, I do understand the reluctance to take on some of the issues associated with terrorism. Terrorism is a phenomenon propagated by people, against people. It is tempting to avoid issues that may be offensive to persons of any particular sociocultural or socioreligious group. We live in an age where political correctness is often substituted in place of accuracy and fact. Furthermore, when a person is dealing with issues concerning life and death for another person, there is a tendency to be conservative with the information so as to avoid giving too much advice when the consequences can be so grave. Indeed, these are legitimate and heavy concerns.

But I have other concerns as well. The first is in attempting to avoid any possibility of offending any group of people, we tend to avoid topics that deal with the actions or physical appearance of another person. By doing so, I believe a powerful and effective technique used to identify potential threats, called profiling, is often avoided completely. My second concern is that when we intentionally limit the information given a person, we unintentionally limit their options for survival. It is precisely when the consequences are so grave and the outcome so uncertain that the maximum amount of information would be the most helpful. The individual who has inadvertently become the target of terror must ultimately make the decisions that will save his own life or seal his own fate.

Terrorism is always defined from a point of view. I will be clear so there is no confusion about my point of view. I am a citizen of the United States of America. Mine is the perspective of a person who calls the United States home and who believes this is the greatest country on Earth. I am not interested in seeing an act of terrorism against my fellow Americans from the terrorist's perspective unless this view will help me defeat him and his ability to carry out acts of terror. I will not be drawn into highbrow arguments of why these madmen wish to kill Americans and to destroy the American way of life. Terrorism is a reality that has been brought to our soil and against our people. As an American I will do everything in my power to run the terrorists from our soil and from the dark corners that provide them safe haven. I am not part of the "blame America first" crowd. If America didn't give the fanatics a reason to terrorize our land and people then they would create a reason. It is our success, our liberty, and our freedom that they hate.

In this book I will deal with you, the reader, in a straightforward manner. I will treat you with respect and assume that you are an intelligent, thoughtful person who is concerned with your own life and the lives of others. I will give you the information I believe could be important to you during these most extreme atrocities. I will not tiptoe around topics I believe are essential to your survival just to avoid the penalty of stepping out of the "political correctness" bounds. I will not tell you that every situation has a way out because that is not reality. I will tell you that there may be some hard decisions to make, and some of those decisions could end your life or the life of another. This book is lean and you won't find filler or fluff in its pages. This is a serious book to be read by those persons serious about preparing for, responding to, and surviving terrorist attacks. Most of all, I hope you will never need to apply the knowledge contained between the covers of this book.

Brett Earnest
San Francisco, USA
July 4, 2003

What Is Terrorism?

Defining terrorism is a slippery and politically charged task. Even when a definition is proposed it sometimes fails completely when it is applied to the messy reality of our geopolitical world. Most definitions of terrorism that I have encountered in my study for this book have left me still wanting more clarity. So, having found no satisfactory definition of terrorism, I have been left to my own devices to create a workable definition of terrorism for the narrow scope of this book. I am certainly no politician and only a dubious scholar in the eyes of those highly esteemed academics who prowl the halls of our lofty institutions. Having said that, I will offer my definition with the understanding that it, too, may very well fall short in one way or another.

For the purposes of this book, I shall define terrorism as the real and tangible planning, support, and execution of acts of violence against any target, military or civilian, when that target reasonably should assume that no state of war or hostile intent exists within the context of its environment at the time of attack. Furthermore, an act of violence is considered to be any act carried out with the intent of killing, injuring, striking fear in, or disrupting critical services to the general population.

This seems like a terribly wordy definition, but I want the definition to accurately embody terrorism in the real world. It is easy to see that the attack on Pan Am Flight 103 over Lockerbie, Scotland in 1988 was an act of terror. Flight 103 was a civilian aircraft carrying civilians during the holiday season when it was blown from the sky by a bomb. Less clear, however, is the attack on the United States Marine barracks in Beirut, Lebanon in 1983. The Marines were in a hostile environment in a peacekeeping role trying to bring order to the chaos in Lebanon. But when you consider the violent act in light of the definition I have provided, you will see that the Marines were victims of a terror attack because of the *context* of the attack. They were not there as a conquering or occupying force, but rather were one part of a multinational force dispatched to keep peace in the region and were welcomed guests of the government of Lebanon. As I said before, the definition of terrorism is a slippery one in the real world.

Selected Cases Of Terrorism Against Americans Overseas

The following selected cases were chosen from an extremely long list of attacks on Americans and American interests overseas. It is obviously a short listing and not exhaustive by any means. These cases have been selected because of their generally high recognition factor and for the purpose of pointing out a more sinister development over the past 20 years or so.

In reviewing these cases you will notice we begin with a rather impromptu mob of students storming an American embassy in Tehran, Iran. The Tehran mob-style, unfunded, raid of the American embassy lies in stark contrast to the very well planned, funded, and executed attacks on the World Trade Center and Pentagon in 2001. The progression from disorderly street scenes to highly organized attacks is strikingly clear. It shows the terrorist organizations around the world have become increasingly better organized and now have access to millions of dollars to support their activities. It is also obvious that Americans and the United States government are in dire need of developing, funding, and maintaining intelligence efforts to keep track of individuals and organizations that pose a threat to Americans. Along with the intelligence efforts must come the willingness and ability to use decisive force when the need and opportunity arises.

THE IRAN HOSTAGE CRISIS

In the early months of 1979 the rule of the Shah of Iran came to an end amid an uprising of Islamic fundamentalism led by the Ayatollah Khomeini. The Ayatollah had been exiled from Iran the year before and was living in Paris, France at the time of the uprising. The Shah of Iran had a good relationship with the United States, but his rule had become increasingly marred by bloody confrontations with revolutionary factions within the country. The relationship between the United States and the government of the Shah of Iran had created an anti-American sentiment in Iran. Relations between the United States and Iran became strained under the rule of Ayatollah Khomeini culminating in a hostage crisis that lasted for more than a year.

On November 4, 1979 Islamic students in the capital city of Tehran stormed the United States embassy and took ninety hostages. Soon thereafter all non-American, women, and black hostages were released, leaving over fifty Americans as hostages of the Islamic militants. Lengthy negotiations by President Jimmy Carter proved useless as the hostage crises worsened. In April of 1980 a rescue attempt failed before it really got underway, resulting in the deaths of eight United States servicemen. Then, on September 22, 1980 Iraq invaded Iran, leading to an all-out war between the two countries and further complicating negotiations to free the hostages. Finally on January 20, 1981, a full 444 days after their capture, the hostages were released as President Ronald Reagan took office in the United States.

MARINE BARRACKS: BEIRUT, LEBANON

The government of Lebanon was embroiled in a tumultuous religious and political struggle in 1983. As guests of the Lebanese government, a multinational force was called into action and dispatched around the city of Beirut for peacekeeping purposes. The United States Marine Corps 24th Marine Amphibious Unit was dispatched to Beirut by President Ronald Reagan and they set up barracks in a four-story concrete building at the Beirut airport.

Several governments and factions were in conflict in Beirut. The Israelis, Syrians, and the Palestine Liberation Organization saw Beirut as a strategic piece of the geographic puzzle that makes up the Middle East. For the most part, the Marines had a good relationship with the local population. Occasionally, however, the Marines had the difficult job of being involved in local skirmishes in order to keep the peace. Theirs was a mission of peace in Lebanon.

Just after 6:00 AM on October 23, 1983 a Muslim extremist driving a Mercedes truck loaded with explosives drove into the first floor of the Marine Corps barracks and detonated the charges. The resulting explosion ripped the concrete building apart and in the process killed over 240 Marines and other servicemen. It was the largest single-day loss of Marines since the battle for control of the Japanese island of Iwo Jima in World War II.

TWA FLIGHT 847

On June 24, 1985, two Shi'ite Muslim terrorists hijacked TWA Flight 847 and began one of the most bizarre series of events ever to be experienced in hijacking history. The flight had originated in Cairo, Egypt and made a stop in Athens, Greece. The terrorists took control of the aircraft about thirty minutes after it left Athens en route for Rome, Italy. On board were 153 passengers and crew; around 100 of them were Americans.

The terrorists demanded that the pilot land the aircraft at the airport in Beirut, Lebanon. Beirut authorities at first refused permission for the terrorists to land, but threats of violence against the passengers persuaded the authorities to allow the aircraft to land. Fuel was demanded as passengers were beaten and threatened with death. The aircraft was refueled, a few hostages were released, and the aircraft took off from Beirut en route to Algiers, Algeria.

The Algerian government turned off the airport lights in an attempt to keep the hijacked aircraft from landing. A call from United States President Ronald Reagan to Algerian President Chadli Bendjedid resulted in the opening of the airport to the hijacked aircraft. It landed in Algiers and the terrorists released several more hostages before ordering the pilot once again to leave Algiers and return to Beirut.

After landing in Beirut a second time another dispute over refueling the aircraft took place. The terrorists took American passenger Robert Stetham, who was a United States Navy diver, to the front of the aircraft. There they beat Stetham severely then shot him in the head, killing him instantly. His body was thrown from the aircraft and the terrorists exclaimed on the radio that everyone should now take them seriously. Shortly thereafter officials from Amal, a Shi'ite Muslim organization with ties to Syria, entered the aircraft to negotiate with the terrorists. The terrorists demanded that the lights of the airport be turned off. While the lights were off at least ten more terrorists were brought on board the aircraft as reinforcements to the two terrorists that had hijacked the aircraft the day before. Also in the darkness the terrorists removed six American citizens with military identification or Jewish sounding names from the aircraft and took them into the neighborhoods north of Beirut. The terrorists made demands including the release of Shi'ite Muslims being held by Israel, the withdrawal of Arab money from western banks, and the release of more Shi'ite prisoners in Kuwait and Cyprus. After stating their demands and another refueling the pilot of the aircraft was ordered to fly once again to Algiers, Algeria.

More passengers were released upon arrival in Algiers, leaving less than forty of the original passengers on the aircraft. All of the remaining passengers were American men. Within hours the aircraft was on its way to Beirut for a third time. The Beirut authorities again refused the pilot the right to land at the airport. After a heated exchange the pilot declared he had only five minutes of fuel left and he landed the aircraft at the Beirut airport. Under cover of darkness Shi'ite Muslims removed the remaining hostages from the aircraft and took them to various locations in Southern Lebanon. The terrorists blew up the aircraft and fled into Southern Lebanon as well.

Over the next several days the United States sent in Delta Force units to rescue the Jewish and military passengers that had been removed from the aircraft during the second stop in Beirut. The rescue operations were successful. The remaining hostages held in Southern Lebanon were under the control of the Amal and Hezbollah militias. The freedom of the remaining hostages was secured by Syrian President Hafiz al-Assad. The terrorists were never caught.

THE ACHILLE LAURO

On October 7, 1985 a group of terrorists associated with the Palestine Liberation Front (PLF) hijacked the luxury cruise liner Achille Lauro in Egyptian waters. Around four hundred passengers, mostly elderly tourists, were taken hostage by the terrorists. The terrorists communicated with the Egyptian authorities telling them of their demands that Israel release fifty Palestinian prisoners being held by the Israelis. The captain of the Achille Lauro was ordered to sail the ship to Tartus, Syria. Upon arrival the next day the ship was denied docking rights by Syria. After receiving the news of the denial to dock the terrorists took Leon Klinghoffer, an elderly, wheelchair-bound, American passenger to the edge of the ship. There the leader of the terrorists shot Leon Klinghoffer in the head and chest and pushed him, in his wheelchair, over the side of the ship into the water below.

Still the Syrians denied the Achille Lauro the rights to dock at Tartus. Threats of a second murder were made, but a radio message from the PLF leaders ordered the leader of the terrorists aboard the Achille Lauro to leave the passengers unharmed. They were ordered by the PLF leaders to set a course for Port Said, Egypt. Upon arrival and docking at Port Said, the Egyptian government negotiated the safe release of all hostages in exchange for a guarantee of safe passage for the terrorists. At the time, the Egyptian government was unaware that the terrorists had murdered an American citizen.

The release of the hostages brought to light the murder of 69 year-old Leon Klinghoffer. The terrorists were placed on an EgyptAir charter flight bound for Tunis, Tunisia despite objections from the United States. President Ronald Reagan ordered the jet carrying the terrorists to be intercepted by F-14 Tomcat fighter jets dispatched from the USS Saratoga. The EgyptAir jet was diverted to a NATO base at Sigonella, Sicily where it landed. The jet was immediately surrounded by American troops. The American troops, in turn, found themselves surrounded by Italian troops. Negotiations between Ronald Reagan and Prime Minister Benitto Craxi of Italy resulted in the handing over of the terrorists to Italy.

It was found that the terrorist leader, Abu Abbas, had been in direct communication with the Palestine Liberation Organization leader Yasser Arafat. In speaking with Prime Minister Craxi, Arafat warned of retribution by militants if Abbas was handed over to the Americans. In the end Italy refused to give the terrorists to the United States. Abu Abbas was released because the Italian authorities claimed that they didn't have enough evidence to hold him. All four terrorists were convicted in an Italian court and sentenced to long prison terms. Abu Abbas, however, was sentenced in absentia but never served prison time for the terrorist attack on the Achille Lauro. In April of 2003 after the war on Iraq led by the United States and Britain, Abu Abbas was captured near Baghdad, Iraq. His disposition at the time of this writing is still being debated.

PAN AM FLIGHT 103

On December 21, 1988 Pan Am Flight 103 taxied down the runway at London's Heathrow International Airport with 243 passengers and 16 crewmembers aboard. Of those aboard, 189 were Americans. The flight took off at 6:25 PM local time and began climbing to its cruising altitude for the transatlantic flight to New York. Thirty-eight minutes later at an altitude of 31,000 feet over Lockerbie, Scotland, Flight 103 exploded and fell from the sky.

The flaming wreckage covered a huge area and many houses on the ground were set ablaze. In Lockerbie, eleven people lost their lives as pieces of the wreckage crashed into houses and fields below. In all, 270 people died in the explosion and wreckage of Pan Am Flight 103.

An extensive investigation eventually tracked the source of the explosion to a Toshiba radio-cassette player that had been packed with Semtex plastic explosive and a timing device in a brown suitcase. Eventually two Libyans, Lameen Fhima and Abdelbasset Megrahi, were charged with planting the device. Various specific reasons have been proposed as to why the Libyans chose to attack the aircraft, but relations between Libya and the United States had been openly hostile for several years.

The two Libyans were brought to trial in 2001 before a special Scottish Court in Camp Zeist, The Netherlands. Lameen Fhima was acquitted of all charges while Abdelbasset Megrahi was found guilty. Megrahi was sentenced to life in prison with the possibility of parole in 20 years.

KHOBAR TOWERS

On the evening of June 25, 1996 a tanker truck pulled into the front area of Khobar Towers in the city of Dhahran, Saudi Arabia. The tanker came to a stop about 80 feet in front of Building 131, which housed about 100 United States Air Force personnel. Sentries on the roof saw the tanker and thought it was suspicious. They immediately began evacuation procedures, but their attempts were cut short when the 5,000 pounds of plastic explosives packed inside the truck was detonated. The resulting explosion cut off the entire face of Building 131 and killed 19 Americans while wounding over 300 people in the surrounding complex. The bomb was more than twice as powerful as the truck bomb used in the terrorist attack on the Marine barracks in Beirut, Lebanon in 1983.

An investigation led investigators to the Islamic militant terrorist organization Hezbollah Al-Hijaz. This terrorist organization had been banned from Saudi Arabia for some time, but its operatives continued to meet in the surrounding countries of Syria, Lebanon, and Iran. Hezbollah Al-Hijaz promoted and instigated violence against Americans on Saudi Arabian territory and its leadership was claimed to have close ties with officials in Iran.

In June of 2001 fourteen terrorists were indicted by the United States for their involvement in the Khobar Towers attack.

KENYA AND TANZANIA EMBASSIES

On Friday, August 7, 1988 two nearly simultaneous attacks on the American embassies in Nairobi, Kenya and Dar es Salaam, Tanzania killed over two hundred people. Of the victims killed, twelve were American. All of the Americans killed were in or near the Nairobi embassy.

The attacks were well coordinated and meticulously planned. This was to become the mark of a radical Muslim terrorist organization called al Qaeda. The name al Qaeda means "the base". Al Qaeda was started, financed, and led by a Saudi Arabian exile named Osama bin Laden.

THE USS COLE

On October 12, 2000 the USS Cole, a 505-foot guided missile destroyer, was refueling in Aden Harbor, Yemen. Part of a routine stop made by United States Navy ships patrolling the region, the USS Cole was operating under threat level "Bravo". The lowest threat condition is Alpha; Bravo is next highest; Charlie is higher; and Delta is the highest. Threat condition Bravo was the normal condition for refueling at this Yemeni harbor.

The sailors aboard the USS Cole took notice of a small boat headed toward the billion-dollar destroyer, but did not stop the boat from approaching. Within minutes the small boat had reached the USS Cole and pulled close alongside. Without warning, the two men on the small boat saluted and the boat exploded with incredible force. A gaping forty-foot hole was torn into the hull of the USS Cole, nearly sinking her. Of the crew of the USS Cole, 17 were killed and another 39 were injured. An investigation revealed that the radical Muslim terrorist organization al Qaeda, led by Osama bin Laden, had been responsible for the attack.

DANIEL PEARL

Daniel Pearl was a graduate of Stanford University having majored in communications. He had been a respected journalist for fourteen years. As a journalist for the Wall Street Journal, he had lived and worked all over the world and was well respected among his colleagues.

After the attacks on America by al Qaeda on September 11, 2001, Daniel Pearl opted to write about militant Islamic groups. He was on this assignment in Karachi, Pakistan when he was kidnapped on January 23, 2002. Pearl's kidnapping was carried out by an organization calling itself the National Movement for the Restoration of Pakistani Sovereignty. His captors had lured Pearl to a meeting where he was investigating a possible connection between would-be shoe bomber Richard Reid and Islamic militants in Pakistan. Having captured Pearl, his captors interviewed him and videotaped the interview for later use. Pearl's captors eventually beheaded him and the beheading was also videotaped. A video collage was created by the militants and sent to officials in Pakistan. Daniel Pearl was 38 years old and his wife, Mariane, was seven months pregnant at the time of his murder.

Selected Cases Of Terrorism In America

For years the United States enjoyed the safety afforded us by the ocean borders to our east and west, and friendly relations to our north and south. Terrorism was something that happened "over there" and was largely seen as a foreign affairs problem. We had become complacent to the attacks happening in other parts of the world, even though those targets were Americans and American interests.

On September 11, 2001 the United States of America got a kick in its complacency. We were awakened to the sound of roaring jet engines and crashing buildings, and our eyes were finally opened to the threat that had been gathering for years. Terrorism had burst onto American soil from over the waters and across the borders that had seemed so insurmountable, so friendly only days before. The warning signs were there, but connecting the scattered dots with fund-strapped agencies that had seen their effectiveness eroded by partisan politics was quite impossible. The fact is that the deterioration of the intelligence community brought about by the "doves" in government who didn't have a stomach for the dirty work of underworld spying had left us wide open to the horror that befell our land. The victims of 9-11 paid the ultimate price for the lack of leadership and determination exhibited in government. We will be paying this price for years to come because our enemies have become emboldened and we have a lot of catching up to do.

The following selected cases are a scattering of incidents that range from terrorists that originated from outside our borders to terrorism from our own citizens. On the positive side of these attacks is the fact that there have been so few, though some were devastating. On the negative side of this short list of terrorist incidents within our borders is the reality that this list will become longer with time. It is my sincere hope that we do not fall into the doldrums of complacency again.

THE DALLES, OREGON

In 1981 a self-proclaimed guru, Bhagwan Shree Rajneesh, and many of his fol-
lowers moved from India to a newly purchased 64,000 acre ranch just outside of
Antelope, Oregon. The Rajneeshees, as they came to be called, immediately set
up an impressive small city on the ranch. By 1984 the city boasted a large airstrip,
shopping facilities, a huge meeting hall, water reservoir, and a new, sophisticated
medical facility.

All was not well, however, and the cultural differences between the
Rajneeshees and the local people caused a lot of friction. In addition, the
Rajneeshees were building without the necessary permits, those having been
denied by the Antelope city council. With their numbers strong in the district,
the Rajneeshees were able to win council seats in Antelope. After the election of
several cult members to city council, the city of Antelope, Oregon was renamed
to City of Rajneesh, Oregon.

The City of Rajneesh existed within Wasco County, and the county seat of
Wasco County is the city of The Dalles, Oregon. Desiring more influence countywide, the same tactic of taking over elected offices that they had used in Antelope was on the mind of the Rajneeshees. There was a problem in this plan for
the Rajneeshees because countywide they did not have enough of the voter population supporting them. To remedy this, the Rajneeshees used their medical facility to culture strains of *Salmonella typhimurim*. The Salmonella was tested a
couple of times, once by contaminating the produce section of a local supermar-
ket, and another time by directly lacing the water given to two council members
from The Dalles. Several dozen people became sick from the biological agent that
had been distributed. Finally near the end of September a larger, much more devastating attack took place when the Rajneeshees contaminated the salad bars,
creamers, and dressings of ten local restaurants. They also smeared their biological agent on the doorknobs and urinal handles in the courthouse in The Dalles.
In the end over 750 people were attacked and sickened by the biological agent,
but there were no deaths.

The success of the attack brought close scrutiny of the local restaurants by
health organizations. Officials from all over the state swarmed The Dalles in an
attempt to locate the source of the agent. The presence of so many officials
helped prevent the more massive attack planned for the next month during the
election period. It took more than two years and the defection of several former
Rajneeshees to uncover the truth. The truth was that The Dalles had been
attacked using a biological agent and with massive effect. It was also revealed that

the Rajneeshees possessed the precursors to typhoid fever, and had cultures of tularemia as well. It could have been much, much worse.

THE WORLD TRADE CENTER BOMBING

The World Trade Center in New York City came under attack on February 26, 1993. The underground parking garage of the North Tower was shaken by a powerful explosion, the result of the detonation of more than one thousand pounds of explosives. The explosives were packed into a truck and the truck was parked in a position that the terrorists believed would cause the North Tower to fall into the South Tower and in turn collapse the South Tower, causing both to crash down upon the surrounding buildings. Along with the explosives were mechanisms to disperse cyanide gas. If all had gone according to plan, the death toll would almost certainly have reached tens of thousands.

As it happened, however, the North Tower stood, the South Tower was left intact, and the explosion itself burned up the cyanide gas before it had a chance to harm anyone. Even so, six Americans died and over a thousand were injured. A three-story crater was blown into the base of the North Tower and caused nearly five hundred million dollars in damage.

Among the terrorists responsible for the attack were Ramzi Yousef and Abdul Rahman. All of the terrorists eventually found to have been responsible for the bombing were Islamic militants. Ramzi Yousef has been suspected of having relations with Iraqi intelligence. Ramzi Yousef fled the United States. In January 1995 Yousef and a few associates were planning multiple, simultaneous bombings of eleven jet airliners over the Pacific Ocean. Their plan was to use liquid explosives in packages to avoid detection as they passed through the airports. While mixing the volatile liquids in his apartment in Manila an accident started a fire that forced Yousef to flee. In the process, Yousef left behind a computer that eventually led to his arrest in Pakistan. He was extradited to the United States for prosecution.

THE MURRAH FEDERAL BUILDING

At 9:02 AM on April 19, 1995 a powerful explosion sheared off the face of the Alfred P. Murrah Federal Building in Oklahoma City, Oklahoma. The blast killed 168 people and injured several hundred. Of the 168 victims, 19 of them were children.

The bomb was made up of two tons of fertilizer mixed with nitromethane racing fuel. The slurry was placed in containers and loaded into a 20-foot rental truck. The truck was parked just outside of the Murrah Federal Building. At 9:02 AM the bomb detonated and the front of the building collapsed. Though made with readily available materials and easily obtained knowledge the bomb was devastating.

Timothy McVeigh was accused and convicted of the bombing. An accomplice, Terry Nichols, was tried and convicted as well. McVeigh was sentenced to death and on June 11, 2001 he was executed. Terry Nichols was sentenced to life in prison.

Though McVeigh took his motivation for the attacks with him to his grave, there is widespread speculation that the Murrah Federal Building attack was in retaliation for the attack on the Branch Dividian compound in Waco, Texas. Further speculation has been that the attack was also in retaliation for the United States government's attack on Randy Weaver's family at Ruby Ridge, Idaho in 1992, which resulted in the death of Randy Weaver's dog, 14-year-old son, and wife. Both incidents involved agents working out of the Murrah Federal Building in Oklahoma City.

The reason the Oklahoma City attack is important is that it is an example of domestic terrorism. It was a low-tech weapon delivered in a low-tech manner by one person operating without the support of a large network. Nevertheless, the results were catastrophic. Terrorism can be brought to us from outside our borders but may also be a result of homegrown fanaticism. The latter is the case in the bombing of the Murrah Federal Building.

SEPTEMBER 11, 2001

The morning broke bright and clear on September 11, 2001. No Americans rising that morning knew the horror that would befall us before lunch was served on the east coast.

At 7:59 AM American Airlines Flight 11 took off from Logan International Airport in Boston, Massachusetts with 92 passengers and crew aboard. Flight 11 was bound for Los Angeles, California.

At 8:01 AM United Airlines Flight 93 took off from Newark International Airport in Newark, New Jersey with 45 passengers and crew aboard. Flight 93 was bound for San Francisco, California.

At 8:10 AM American Airlines Flight 77 took off from Washington Dulles Airport in Washington, DC with 64 passengers and crew aboard. Flight 77 was bound for Los Angeles, California.

At 8:14 AM United Airlines Flight 175 took off from Logan International Airport with 65 passengers and crew aboard. Flight 175 was bound for Los Angeles, California.

In fifteen minutes all flights were in the air, as were 19 terrorists scattered among the passengers on all of these flights. Each flight had terrorists trained to fly the aircraft while other terrorists had the responsibility of controlling the passengers. Soon after the flights were in the air the terrorists went to work.

Using knives and razor-sharp box cutters the terrorists took control of each aircraft by cutting and stabbing passengers. They gained access to the cockpits of each aircraft through intimidation and brute force. Never knowing the sinister plans of these terrorists, the passengers at first did what seemed reasonable based upon all other hijackings that had taken place in history. They probably reasoned that they should bide their time and allow professionals to deal with the terrorists when they landed. At the time, this was indeed the most reasonable thing anyone on any of these flights could have imagined.

At 8:28 American Airlines Flight 11 abruptly turned off course and the transponder that identifies the aircraft to air traffic controllers was turned off. Flight 11 became a blip on the radar, and that blip was heading for New York instead of Los Angeles. A few moments later Flight 175 turned around from its westward course and made a heading for New York. Indications were that something was also terribly wrong aboard Flights 77 and 93.

At 8:45 AM Flight 11 struck the North Tower at the 100th floor level. At the time many believed that it was an accident but 18 minutes later, at 9:03, Flight 175 plunged into the South Tower at the 90th floor and everyone across the country realized America was under attack.

With the World Trade Center ablaze, people evacuating, and firefighters running up the flights of stairs to fight the inferno another airliner found its mark over 200 miles away. Flight 77 plowed directly into the west side of the Pentagon in Washington, DC. Heroic efforts were made to bring people out of the rubble before the floors collapsed. The west side of the Pentagon crumbled just after 10:00 AM.

Meanwhile, amid frantic cell phone calls from passengers aboard Flight 93, the stage was set for the first soldiers of the new war on terrorism to become heroes. Knowing the fate of the other hijacked airliners and surmising that they were aboard a weapon to be used against America, the passengers of Flight 93 bound for San Francisco apparently attacked the hijackers. The aircraft flew erratically in the Pennsylvania skies and finally was flown into the ground in a Pennsylvania field. All passengers aboard the aircraft were killed, but they undoubtedly saved countless others by denying the terrorists the opportunity to use the aircraft as a weapon against fellow Americans.

It is my opinion that if the passengers of Flights 11, 77, and 175 had known the intent of the terrorists they, too, would have attacked the terrorists and flown them into the ground. It was by chance that they were the first to be taken unaware and their actions were reasonable given the information they had. In death they served as a warning to the passengers of Flight 93. The passengers of Flight 93 took that information and aborted the mission of the terrorists, making the ultimate sacrifice in the process. These are Americans at their best.

Subsequent investigation determined that the leader and financier of the al Qaeda terrorist organization, Osama bin Laden, was responsible for the attacks. In the end over 3,000 people were killed in this attack on America by Muslim extremists.

If one lesson can be learned from September 11, 2001, it is that terrorists are organized, committed, well financed, and fanatical in their hatred for the United States of America. We, too, must be organized, committed, well financed, and fanatical in grinding them and their organizations into the dirt.

THE ANTHRAX ATTACKS

During the last three months of 2001 several letters were sent through the United States mail system that were laced with the biological agent anthrax. The letters containing anthrax were sent to media corporations and government offices up and down the east coast. The letters contained notes acknowledging the deadly contents. Further analysis confirmed that the strains of anthrax were no more than a couple of years old, possibly indicating that the person who had gained access to the anthrax may have worked in a laboratory or industry that deals with this sort of biological agent. At the time of this writing, the person responsible for the letters has not been apprehended and may still have access to anthrax in the future.

The contaminated letters had been circulated through the mail system and inside offices where people came into contact with the anthrax agent. Of the many who were exposed, 23 people became ill and 6 of them died. The anthrax victims had both skin and inhaled forms of the infection. Skin anthrax infection is not nearly as serious as the inhaled form. Both, however, respond to antibiotics when the infection is caught early. Because of this, antibiotics were given to those who actually were infected with anthrax as well as those who had been in areas suspected of anthrax contamination.

The cost of cleaning up the anthrax was measured in millions of dollars, even though this was a low-tech method of distribution involving only a few ounces of anthrax agent. The fear factor of knowing that the mail system had been used as a distribution method for the agent put the whole nation on edge. As devastating as the casualties were, the casualties were few. The widespread disruption the anthrax attack caused far exceeded the actual risk posed to any one individual. This type of asymmetric reaction is the ultimate goal of terrorism, and distributed agents are the perfect tools for producing this result.

Terrorist Organizations

Terrorist organizations are those affiliations that exist for the purpose of providing monetary, materiel, and logistical support of individuals or groups of individuals that engage in terrorism (as defined earlier). Some terrorist organizations are longer-lived than others. Organizations that have religious or ideological differences with the United States and have funding and logistical support from states or wealthy individuals have the capability of lasting for decades. Other terrorist organizations that are more concerned with United States policy and lack consistent financial backing may come and go quickly.

Terrorist organizations have provided the United States with a formidable problem of identifying and locating their bases of operation. Tracing the trail that leads to their monetary support is a lengthy and arduous task as well. Terrorist organizations do not have borders and some, such as al Qaeda, have terrorist training camps scattered in many countries across the globe. Several of the countries where terrorists train and congregate either actively support the terrorist organizations or at least do not interfere with their activities. By using the cover of the host country, terrorist organizations further complicate any effort to remove the facilities and means used by them. An attack on a terrorist camp that lies within the borders of a sovereign nation can be seen as an act of aggression, or even an act of war, by the terrorist organization's host nation. Knowing this, terrorist organizations take advantage of any piece of ground that affords them the ability to operate, train, and launch attacks without interruption from outside forces. Many governments that are sympathetic to the cause of terrorists may be loosely linked by a common hatred for the United States and a common religious conviction. This particular combination has proven to be a particularly strong relationship. In its most advanced form, the line between the host country or government is blurred with the terrorist organization itself. Such was the case in Afghanistan where the Taliban government and the al Qaeda terrorist organization were so tightly bonded that they fought side-by-side against United States and Northern Alliance forces.

Cooperation between host countries and terrorist organizations can blur the view within the international community as well. Differing religious, cultural, and political views across the globe allow for varying levels of tolerance for nations that host or indirectly support terrorist organizations. One nation that has significant economic ties with another nation that is a host or supporter of terrorism may find itself even more tolerant of otherwise unacceptable behavior. The difference between stage and actor, host nation and guest become nearly indistinguishable on the international level. Bringing to focus the differences and agreeing on the proper remedy is a nearly impossible task. International organizations could be effective in dealing with terror on a global scale if they would speak with one voice, but this is rarely the case. Bitter disputes and political infighting have made most international consortiums ineffective in dealing with terror. Ultimately, each nation must accept the responsibility of protecting its citizens from terrorism. Only by identifying the terrorists and their political, materiel, and monetary support structures can the nations of the world strike at the heart of the terrorists who pose a threat to their citizens. Indeed, it is the responsibility of all free and peace-loving nations to do so.

Types Of Attacks

In their infinite imagination, terrorists have devised countless ways to maim, kill, and disrupt the lives of free people all over the world. Still, these various destructive means can be roughly grouped into categories. I will discuss four major categories throughout the following chapters. For now, I will briefly describe each of these categories in preparation for a more comprehensive treatment of each later.

INFRASTRUCTURE ATTACKS

Infrastructure attacks are those attacks in which the victim is harmed as a result of actions taken against societal infrastructures. Some examples of infrastructure attacks would be computer-based (cyber) attacks against power grids, hospitals, mass transit, air and land traffic control, and economic targets. In this book I will offer some suggestions to aid you in preparing yourself for surviving without services that may be interrupted during an infrastructure attack.

DIRECT ATTACKS

Direct attacks are those attacks in which the victim is harmed as a result of direct contact with the terrorist. Some examples of direct attacks would be shootings, hijackings, and physical assault by the terrorists against you or those around you.

In some cases direct attacks may have the purpose of wounding and killing as many people as possible with the initial attack, such as an attack by assault rifle or handgun. In other cases the main mission of a direct attack may be to take control of a vehicle, such as a train or aircraft, in order to use it as a weapon to inflict as much damage or cause as many deaths as possible. In this book I will discuss different scenarios that may help you to avoid, prevent, resist, and survive direct terrorist attacks.

DISTRIBUTED AGENT ATTACKS

Distributed agent attacks are those attacks in which the victims are harmed through contact with a substance distributed by terrorists. Some examples of distributed agent attacks would be the distribution of chemical, biological, or radiological agents into a population.

The nature of a distributed agent attack is such that you may be a victim of the attack without immediately being aware that you have been involved in the attack. You could even become an unwilling, unaware facilitator of a distributed agent attack. As an example, you could be infected with the smallpox virus for days during its incubation period before exhibiting any of the symptoms associated with smallpox. As the telltale rash begins to develop, you are steadily becoming more infectious. During this time, you are spreading the virus to those you come into contact with in your family, on the bus, and at work. By the time you are diagnosed with smallpox, many of those with whom you have come into contact will have already become infected (and by now could have exposed many of their friends and family to the virus as well). In this book I will give you some suggestions on avoiding and surviving distributed agent attacks.

EXPLOSIVES ATTACKS

Explosives attacks are those attacks in which devices are detonated with the intent of causing casualties with the initial blast and shrapnel, or by the failure of a building or structure damaged by the blast. These attacks include suicide bombers (also called homicide bombers), remotely triggered explosions, timed bombs, booby traps, car and truck bombs, and nuclear weapons. These types of attacks are very difficult to defend against because they can happen with no warning and usually pack an incredible, shocking force. With relatively little knowledge and limited financial layout, a terrorist can wreak havoc by detonating a homemade device with devastating impact. In this book I will discuss explosives attacks and some steps you may take to reduce your exposure to these types of attacks.

Types Of Targets

Serious terrorist organizations are not haphazard in selecting targets to attack. The targets the terrorists choose usually have a particular audience in mind and a specific message to be sent. Terrorists may spend enormous amounts of money and time studying high profile targets, sometimes scouting the targets three to five years in advance of the actual attack. The terrorists will be looking for patterns of security, structural weaknesses, traffic flow, and other items that will allow them to choose the method, location, and time to strike. Terrorist organizations will be looking for the biggest impact for the least amount of financial outlay and risk of failure. There is no reason to attack a difficult target when attacking an easy target can send the same message. In this chapter I will discuss basic types of targets and how they might be categorized in terms of vulnerability to attack.

SOFT TARGETS

Soft targets are buildings, structures, or areas without special barriers, law enforcement, or military protection. Soft targets are usually associated with civilian activities that would otherwise generally be off-limits to an attack by a professional military in modern times. But for terrorists, soft targets represent a low-risk opportunity to cause mass death and incite fear. Some typical examples of soft targets are shopping malls, mass transportation vehicles, crowded city streets and venues, office and apartment buildings, restaurants, dance clubs, and sports arenas.

HARD TARGETS

Hard targets have defensive structures, equipment, personnel, and procedures designed to detect and repel attacks of various types. Hard targets are usually associated with military installations, nuclear plants, and some research facilities.

Some examples of hard targets include Fort Knox, a warship at sea, and Area 51. Obviously, some hard targets are better protected than others.

COMPOSITE TARGETS

Composite targets are a combination of hard targets and soft targets in the same location. Composite targets are usually arranged in layers like an onion, with the soft target forming the outer layer and a hard target at the center with increasingly hardened layers in-between. Federal buildings, NASA Space Centers, and even important people traveling in public are examples of composite targets. If you wish to walk along the crowded sidewalk near the White House, you are free to do so. If you attempt to scale the White House fence (a hardening structural layer) you will be detected by sensors (a hardening equipment layer) and White House security agents (a hardening personnel layer) will be happy to meet you as you come down. If you make it past the first security agents the monitors will be watching and will notify Secret Service agents and Marines (a hardening procedural layer). At first they'll try to tackle you and make an arrest, but as you get closer to the President of the United States, the force will become greater until it becomes lethal to you and your mission. The White House is an example of a composite target.

Areas At Risk Of Terror Attacks

Hard and composite targets have both strengths and vulnerabilities. The types of defenses built around a hard or composite target are designed to protect against particular types of threats. Terrorists will try to find the weakest point in the target defenses and strike there. A terrorist may be able to accomplish the same mission by attacking a soft target. Unlike a formal military, the terrorist is not concerned about innocent lives lost in an attack. To the terrorist, the wanton loss of innocent life is at best irrelevant and at worst, desired.

Terrorists will be looking for vulnerable targets of significance. Targets could include buildings or structures that are symbols of the United States. The World Trade Center was attacked twice, a symbol of the far-reaching economic power of America. The Pentagon and USS Cole were both attacked, symbols of the military might of the United States. Many United States embassies have come under attack, each one a symbol of American political power. The list goes on. Each of these targets had security vulnerabilities and they are symbols of the United States in one form or another.

The general population of Americans or even specific Americans can become the targets of terrorists as well. Americans are the ultimate symbol of Western culture and anything that brings harm to them is seen as a political statement of enormous power.

It would be a good idea for you to take a look around you and try to identify places that may present a vulnerable, symbolic target to the mind of a terrorist. Where do you live? Is there an FBI Field Office next door, or perhaps a nuclear reactor a mile or two down the road? What about your drive to work? Do you travel by a government building or over a bridge that symbolizes America far beyond our borders? Do you work in or near a building that shares floor space with agencies that present a particular thorn in the side of terrorists? Other areas that may present themselves as targets include theme and amusement parks, tourist attractions, mass transit vehicles and tunnels, military posts, government or CIA offices, religious structures, sporting arenas, water or food supplies, corporations, and utilities. The rule here is to examine your environment from the perspective of the terrorist and to identify any areas that would be attractive from that point of view. Remember: the terrorist is looking for an opportunity to make a statement to the world.

Your Greatest Weapon

If I asked you what you consider to be the ultimate weapon for any circumstance, what would you say? I've asked many people this question, and the answers range from knives to guns, clubs to pepper spray, Karate to fleeing. All of those answers are incorrect. The greatest weapon you possess as a human being is your *brain*.

Your brain is the most adaptable tool on the planet. It is faster at processing sight, sound, touch, taste, and smell while simultaneously prioritizing, planning, timing and acting than any computer in existence. Your brain is portable and you'll never leave it at home. Take the time to train your brain to save your life.

MENTAL PREPARATION

One of the most important things you can do before an attack is to prepare yourself mentally. I'm not talking about your feelings, but rather your mental state. To prepare your mental state you must burn into your mind your priorities *before* you are faced with a grave situation. Mental preparedness is a weapon just like a knife or a gun.

When you are faced with a direct attack you are already at a bit of a disadvantage for this reason: the terrorist has a plan. He has already defined his priorities, he has more information about the mission than you do, and he has already gone through the motions of his attacks dozens of times in training for this moment. This will allow him to respond to situations for which he has trained automatically, quickly, and precisely. The terrorist will attempt to immediately seize control of the situation by executing his actions with vigor and determination. This has the tendency to "freeze" the victims in a state of shock momentarily. Sometimes, a moment is all that is needed.

You, on the other hand, have been taken by surprise. You don't know what the plan is, what end result is desired, or what the next move of the terrorist will be. It will naturally take you a moment or two to size up the situation. As you try to assess whether this is the real thing and whether this is really happening to you, the terrorist will be enjoying a huge advantage in tactical knowledge.

In a direct attack, the terrorist may do many things to keep you off-balance mentally. He will first capture the initiative through a seemingly overwhelming use of force, brutality and resolve. His hope is that you will be hesitant for a long enough period of time that he can assume some amount of control over you. He may continue to terrorize you, or he may suddenly become more "reasonable" with you. He may threaten to kill you if you don't do exactly as he says. He may insist that he does not want to hurt anyone, and may instruct you to relax and cooperate so that no harm will come to you. As you can see, the mental battle is already raging. It should now be clear that your mental state is the first battle that you absolutely must win.

Answer as many questions about what your goals would be should you become involved in a terrorist event. A basic knowledge of yourself is all that you will need to answer these questions:

- If faced with a choice, will you attempt to save your life over that of another person? What if the other person is a loved one?

- Will you be able to act alone against a terrorist, or will you need others to support any action you may take?

- Do you have the resolve to seriously injure or kill a terrorist by any means available if the opportunity presents itself?

- Are you prepared to take responsibility if something goes horribly wrong?

These questions are not intended to be a probe of your morality or your feeling of responsibility to your fellow man. These questions are important only for one reason: knowing their answers now will allow you to take decisive and appropriate action in the future. It is important to do some self-searching now so that you will have the answers ready when you are faced with a situation that forces you to make a hard decision. The answers to these questions will naturally sort out some of your priorities. Knowing your priorities will allow you to make fewer, better, and faster decisions in the heat of the moment. The sooner you are able to regain your mental state, the sooner you can begin taking actions that may save your life or the lives of others. You won't have time to think on these things during an attack. This isn't Hollywood, and there won't be time for long good-byes.

I believe you should talk to your spouse or loved ones about what your priorities will be during a terrorist attack. I do not condone throwing your life away needlessly. If you can accomplish your goal and live, then by all means do so. On the other hand, be prepared to do what you must to accomplish your goal. Make your priorities clear to those you care about. If you are a wife or mother, your priority may be your children if they are present during the attack. If you are the husband, your wife and children may be your top priority. Whatever your priorities, you should set them now and commit to them. If you are committed to giving your life to allow your spouse and children a chance to escape, then you should let them know. There won't be time to discuss this during the attack. Say what you need to say now.

Tactically speaking the terrorist is "in the know" about how the events are supposed to unfold. He has practiced responses to what he believes your reactions might be. Still, the terrorist cannot know completely what is going through your mind. Keep it that way, and you will have a weapon that you can turn against him when the moment is right. I will say it again: your brain is your most powerful weapon. Use it!

Environmental Awareness

For our purposes here I will define your environment as the area of your immediate surroundings that you are able to sample with your senses. Seeing, hearing and smelling are primary methods of gathering information about your environment from a distance. Touch and taste are up-close and personal senses. There may be times when you will primarily be using your eyes and ears to scan your environment, while at other times you may be alerted to a potential threat through your nose (as in the case of the release of a toxic substance into the air).

You are always sampling your immediate environment and responding to the information you receive. Your response is adjusted to your expectations about your environment. You might react strongly or you may just ignore the information. Each response is an action, and your response is based upon what you believe to be appropriate considering your environment. If you are walking down the street, you might duck if a baseball is thrown your way. On the other hand, if you are standing on first base wearing a baseball glove, you are going to respond very differently to the baseball headed in your direction.

My purpose in this chapter is to encourage you to take in everything in your environment, and to trust yourself when things don't seem right. Your immediate environment is very often the way in which you will be alerted to potential danger.

INSTINCTS

Before I get into the particulars of what to look for in your environment that may signal a threat to you, I want to mention the use of *instincts*. In a culture that has put so much emphasis on the broad acceptance of unusual behavior, I believe there is a need to reaffirm the value of the hard-wired survival mechanisms that have served humans and animals well for millions of years. For the purpose of this discussion I will refer to these hard-wired mechanisms collectively as *instincts*. The pressure of popular culture to accept, and even delight in, unusual behavior tends to numb our population to the age-old danger alert system called instinct.

I am encouraging you to pay attention to your instincts. Instinct has a raw advantage over culture and logic in sheer speed. Instinct bypasses your "logical filter" and your "acceptance filter" and acts directly to raise your alert level. You may choose to ignore this alert, and in many cases you may be justified in doing so. Ignoring your instincts every time, however, is dangerous and foolish. You may turn off your alert system at your own peril. If something doesn't feel right, there's probably a good reason. Try to identify what that reason is. If you can't find something specific, then you should make a decision to either ignore it or remove yourself from the environment that makes you uneasy. Don't assume your inability to find a specific threat is an indication the threat didn't, or doesn't, exist. Your instinct is a powerful tool in helping you to become aware of threatening persons and events.

PERSONS OF INTEREST AND PROFILING

Persons of interest are individuals that, for one reason or another, have attracted your attention as possible threats to your security. The techniques that allow you to identify a person of interest sometimes come under fire by those who believe that you should never single out a person just because of the way they look, dress, or behave. The word most often associated with the application of these techniques is "profiling". Call it what you will, and think of it as you wish, but profiling is an effective means of filtering out the unlikely threats from the more likely threats.

Profiling is not new and it isn't unique to humans. Profiling is used widely among the animal inhabitants of this planet. Profiling is not the singling out of one particular aspect of a person, but rather takes into account many characteristics and environmental factors. That's what makes profiling so effective. There are many types of profiling including physical, economic, and psychological profiling. Here I will discuss primarily physical profiling because you are generally not in the position to economically or psychologically profile a person in your environment.

Consider an example from the Serengeti Plain in Africa. Many animals live on the Serengeti Plain and migrate across it. Lions and zebras are two of the animals that live on the Plain and they are engaged in a predator-prey relationship. During most of the day, especially in the heat of the day, the zebras and lions sleep and walk within sight of each other without so much as a hint that a life-or-death struggle would ever exist between the two. During this time they largely ignore each other. In the late afternoon and early evening, the adult lions begin to stir and their behavior begins to look less like resting and more like hunting. The zebras, in turn, go to a higher state of alert and begin to watch the lions very closely. As the lions maneuver to find a vulnerable zebra, the zebras herd themselves to minimize individual vulnerability. Sometimes the lions attack first and sometimes the zebras flee first. Either way, the cycle will repeat itself the next day.

The zebras are engaging in profiling based upon their instincts. Notice that the zebras' alert was not triggered just because there was a lion in the area; most of the day the zebra feels free to pass lions at some distance safely. A herd of zebras is not likely to flee from a lion cub, either. Notice, too, that the zebras would not flee if a hawk looked as if it were hunting. Hawks are no threat to zebras. The fact that it is a hawk doing the hunting makes the threat irrelevant to the zebras, so the zebras ignore the hawk. The zebras only respond when the predators fit the threat profile.

The zebras' alarms were triggered in the late afternoon when the adult lions began moving about in a hunting fashion. The keys used by the zebras were environment, age, species, and behavior. This example shows that zebra techniques of profiling are complex and do not rely on any single characteristic. Likewise, your profiling techniques must take into account what the person looks like, how the person is behaving, the age of the person, and the environment in which you find yourself. No single characteristic is a reliable indicator of terrorism or danger, but combined the characteristics can be a strong indicator of intent.

As an individual you have a limited amount of resources for any particular task. You can only do so many things at once. This is the heart of the reason for profiling. In public there are usually more people around you than you are capable of watching effectively. Profiling sets the priorities of your attention so that you can concentrate on persons who are more likely to be a threat while not wasting energy on persons who are not likely to be a threat.

On a recent cross-country flight I connected through a major United States airport. The security policy in this airport was that random checks of carry-on baggage would be made. The tickets were marked (by computer, I think) and the mark indicated whether you would be subjected to a second search before boarding the aircraft. The randomness of the search was guaranteed so that the civil rights of passengers would be left intact. There obviously was a rule against profiling because I, a white male in my forties carrying a computer case, was not stopped. The two twenty-something year old Pakistani men carrying backpacks who boarded the aircraft ahead of me were not stopped either. It was a lady well into her seventies carrying a small purse and holding a walking cane who was stopped. Her shoes were removed and her purse was searched while security guards apologized to her for the inconvenience and cast embarrassed glances at the unbelieving line of passengers. She had been randomly chosen for a second search of her person and belongings. The passengers of the aircraft and the population in general would have been much better served if either I, or one of the Pakistani men on the flight with me, had been searched. Certainly any of us would have been a greater threat to safety and security than the elderly lady subjected to the search. It is exactly this sort of waste of valuable security resources that profiling is designed to avoid.

Profiling is most effective if you know something about the terrorists who may target the people and interests of the United States. It is in your best interest to stay informed about the general appearance and behavior of the terrorists who have been identified by our government. Add to this a little knowledge of normal human behavior and you can work up a profile that could be useful as you try to prioritize the attention you give to people in your environment. Don't dilute your instincts and profiling with what you think might or might not be "politically correct". Have the courage to call it as you see it. When done properly, no one will even know that you are profiling the people around you.

I would suggest that you apply profiling only in concert with instincts. I don't believe it is particularly useful or efficient to constantly profile the people around you when your instincts have not alerted you to a potential threat. There are instances where profiling will alert you to a threat, but most of the time your instincts will be the first to trigger an alert. When your instincts tell you of a potential threat you should begin profiling those persons in your environment. Profiling in this way can be very useful in filtering the likely sources of the threat from those that are unlikely. Profiling acts as a resource management tool when you are actively looking for a specific threat.

EVENTS AND ITEMS OF INTEREST

Events and items of interest are those things that happen around you that attract your attention to a possible threat. An event of interest may be an action taken by a person or it could be something that you see in the absence of a person. For example, a backpack left unattended on your commuter train is an item of interest. It could be a bag of books; it could be a bomb. Either way, it's an item of interest. A van crashes through the gate at a government facility. Maybe the van has lost its brakes; maybe it is loaded with explosives. Either way, it is an event of interest. A man wearing a long jacket on a hot day would be a person of interest. Maybe he's cold natured; maybe he's hiding a machine gun. Either way, he's a person of interest. The point is that anything a little out of the ordinary should be marked as interesting. This doesn't mean that you should immediately take cover. It does mean that your alert level should be raised and you should begin looking for anything else in the environment that will either support or reject the possibility of a threat.

The key to picking up on the persons, events, and items of interest in your environment is to be alert. I don't mean that you should walk around in public scanning every individual you pass from head-to-toe. I'm talking about picking your face up out of your newspaper occasionally and taking a look around. I'm talking about getting your eyes off the sidewalk and up to head level. Look across the street. Cast a look down the alley you are passing. Casually glance behind you. Listen to the sounds around you. The only way to pick up on threats in your environment is to be aware of your environment. Awareness is a major component of staying safe.

Personal Alert Levels

Most of us appreciate order in our lives. It is quite natural for many of us to take a jumbled bunch of items and set about straightening the mess. Humans in general are creatures who tend to categorize everything. The more important a thing is, the more likely we are to have a categorization for it. I want to extend this tendency toward categorization and apply it to the mental state with which you interact with your environment. By knowing what posture you are taking toward your environment mentally at any given time, you will significantly reduce your response time should any threat arise.

I think it is helpful for you to have a Personal Alert Level (PAL) system that is easily defined and easy to recall. Your PAL is a system that categorizes your mental posture with respect to your environment. Your PAL system is broken into four levels, which are represented by colors: green, yellow, orange, and red. These colors are arranged in order, from being under no perceived threat at all to being under imminent threat of attack. I don't think that it is particularly helpful to have more than these four basic alert levels. Too many alert levels tend to confuse matters. I've always been a person who had trouble answering any question that begins, "On a scale of one to ten..." because I never knew exactly what the difference would be between, say, a seven and an eight. For quick recall and ease of use, I think four levels of alert are sufficient and practical. Here are the four levels and an explanation of each.

Green is the level that indicates there is no threat at all. Imagine that you are with your family in the countryside on a picnic. No one is visible for miles around. Maybe you are in your home resting quietly while reading a book or watching television. You could be anywhere that you consider to be safe and is not a likely terrorist target. Your PAL is green. All is well.

Yellow is the level that indicates you are moving about in public, but no person or event has come to your attention as a possible threat. You might be driving a car, riding mass transit, shopping in a mall, walking down the street, or working in your office. You are not completely relaxed, but not on-edge, either. You are aware of your environment and you are actively processing the sights and sounds that are around you. Everything is fine, but there are many sources of information that you are processing. Your PAL is yellow. Stay alert.

Orange is the level that indicates that there is something possibly threatening in your environment. You may have spotted a person, item, or event of interest. Your instincts are telling you that something just doesn't seem right. Your PAL should immediately be raised to orange. When your PAL is orange, you begin to pay very close attention to your environment in an attempt to determine if there are other indicators that an attack could be in the works. Check your surroundings. Where are the exits? Where are the persons or items of interest? Where is the nearest cover? Form a plan and get ready to execute it swiftly and decisively.

Red is the level that indicates an attack is imminent or in progress. Things are rapidly escalating and there may be panic all around. Keep your head. Be decisive and totally committed. Whatever action plan you have made, it's time to execute it with a vengeance. Your PAL is red. There are no rules except survival.

Take some time to get to know your PAL system and use it every day. Just make an occasional mental note about what your current PAL is any time it crosses your mind. Using your PAL system will have the added side benefit of making you pay attention to your environment. Remember that environmental awareness is a major part of staying safe. After a while, knowing your current PAL and what is going on around you will be effortless. You, and the rest of us around you, will be all the safer for it.

Preparing for Infrastructure Attacks

As defined earlier, infrastructure attacks are those attacks in which the victim is harmed as a result of actions taken against societal infrastructures. Responding to an infrastructure attack involves being prepared to do without the services normally supplied by institutions and government agencies. You should take an inventory of the services you have become accustomed to in your daily life and make preparations to do without them for three or more days. Some of the adjustments could be simple while others will require more thought and preparation.

If you are accustomed to using your credit or ATM card for every purchase, you could be in trouble if the banking systems are attacked. A simple electrical outage is sufficient to render credit and ATM machines useless. Keep a small reservoir of cash in an accessible place, preferably on your person.

Be prepared to lose your water services in the event that water utilities are lost in an infrastructure attack. If you are on a well and pump system you could lose access to your water if the power goes out. Ideally you should store at least one gallon per person per day for drinking and basic hygiene. See the chapter *Water Storage, Procurement, and Decontamination* for information about preparing for water emergencies.

Electricity could be cut to your office or home. Electrically operated and controlled mass transit systems may be halted. Keep a flashlight handy so you can move about safely in dark places. Key chain flashlights are lightweight, unobtrusive, and sufficient to light your way. These key chain flashlights may be useful in allowing you to escape or at least reach more substantial lighting. The new LED key chain and full-sized LED flashlights are especially reliable, durable, and they provide light for much longer periods than their incandescent counterparts.

When electricity is lost there may be services and utilities that could stop working and may actually become dangerous for one reason or another. Dams, nuclear facilities, facilities that manufacture chemicals, and other industries may accidentally release substances into the atmosphere that could become hazardous to the population. If you are in the vicinity of such facilities you should be prepared to either leave the vicinity or create an improvised shelter or enter your own ready shelter. A radio or television will be indispensable in this situation and announcements may direct you as to what action you should take.

Power failure will certainly put a strain on traffic signals and other traffic control systems. If you are driving you should stay out of congested traffic areas if at all possible. If you are in a city, you may fare better on foot rather than trying to drive. Gridlock is likely on the roadways. Don't forget water transportation. Boats are not affected by land-based electrical power failure and may prove to be reliable transportation in an emergency. Such was the case on September 11, 2001 after the attack on the World Trade Center in New York. Mariners, along with the U.S. Coast Guard, evacuated thousands of stranded people from Manhattan using various boats and ships. At the time, water transportation was the only way out for thousands of people.

Take a look at your everyday life and determine which activities and items are necessary for you to live and function. Decide which of these are likely to be at risk in an infrastructure attack and make preparations to either do without them or to make them available during an emergency.

Surviving Direct Attacks

As I mentioned previously, direct attacks are those attacks in which the victim is harmed as a result of direct contact with the terrorist. In a direct attack you are in direct physical contact with the terrorist, or are being attacked by an instrument under current and direct control of the terrorist such as a firearm, knife, or club.

A Legal Word

First, I am not a lawyer and I am not limiting my suggestions in this book to those of a lawful nature. What is perceived as lawful in one jurisdiction may not be lawful in another. Also, what is considered moral or just in one part of this beautiful country may draw criticism in another. I am giving my suggestions for action against a terrorist based solely upon what I believe will save your life, and possibly the lives of those around you. At some point, should you ever have to put into action any of the lessons taught in this book, you may be asked to answer for your actions. Your only answer must be that you were in fear of your life and that you were acting in the preservation of yourself and those around you. You should make your own decisions concerning the use of the force I am describing below. The use of force is ultimately your responsibility.

Actions and Consequences

Every action has a consequence. Every thing you do, or do not do, during a terrorist attack is an action. The decision to sit quietly as a hostage is every bit as much an action as deciding to dive out of a window in an attempt to escape. Each decision you make will have a consequence; it is up to you to attempt to make each decision in a manner that will work toward your goal. Set your goal and then proceed to act in a manner that you believe will most likely achieve its end.

CHOOSING TO BE PASSIVE

There are times when it may be wise to be passive during a terrorist attack. Being passive is an action and this action may be the best course to take under the right circumstances. Being passive is not to say that you are giving up and it doesn't mean that you won't combat the terrorists at a later time of your choosing. Being passive is a state that should exist for as long as it suits your purposes and edges you toward your goal.

Being passive in the midst of a terrorist attack likely means that you have been taken hostage or have otherwise been captured by unfriendly forces. In this situation you need to quickly sort out your priorities. One of the major considerations is whether you are primarily responsible for yourself, or whether you are responsible for others around you. You should know that any action you take could influence the chances of survival for those around you. Some actions will increase the chance that loved ones will survive; other actions may seal your family's fate. It's up to you to decide what influence your actions will have.

If you have decided to be passive in a hostage situation you need to be cognizant of the fact that at least some of the hostages will be expendable to the terrorists. Try to keep a low profile at all times. Follow the orders of the terrorists so long as you are not causing harm to other hostages. Avoid making too much eye contact. If a terrorist tells you to get into a position, then get into that position and stay there no matter how uncomfortable it becomes. Never make a move or change position without being told to do so. If you must attract the attention of the terrorist to make a request, then do so by slowly raising your hand. Once recognized, speak slowly and clearly. Your language may not be the native language of the terrorist. Do not yell or become irritated with him. Don't attempt to talk to anyone around you if there is any chance that you will be caught. Do not try to hide belongings or your passport or do anything that will bring attention to you. Terrorists will sometimes choose future victims early in the crisis and use these pre-chosen individuals as needed to make a point. These chosen ones may be beaten or executed if negotiations aren't going the terrorists' way. Don't become a future victim by drawing attention to yourself.

While being passive you should be taking in all of the information you can absorb. Information is power. The more information you have, the better informed you are, and the better decisions you can make. Try to determine how many terrorists are there, what the disposition of each terrorist is, which one seems to be in charge, etc. Identify each to the best of your ability so that you can keep them separate in your mind. Use any distinguishing features that are available to you. If you are keeping your head down and can't see their faces, try to match voices with shoes or clothing. You may even use reflections in windows and polished surfaces to help you see what's going on around you without being too conspicuous. Do whatever is necessary to gain any information that could work to your advantage if there is a need to fight later.

Another reason to be able to identify terrorists is the possibility that in the midst of a rescue, some terrorists may try to escape along with the hostages. Being able to identify the terrorists to the liberating forces could be crucial. The terrorist may only be feigning cooperation in order to attack when the liberating forces let their guard down.

If you are a hostage and an attempt is made to free you using force, you must be prepared to get down in a hurry. At the first sign of a fight between liberating forces and the terrorists you should hit the deck and stay there until instructed to do otherwise. Don't stand up and don't try to run. You will certainly become the target of enemy and friendly fire. Don't try to help the liberating forces by getting into the fight. Let them do their job and stay out of the way. They are professionals and the last thing they need is your help in the heat of battle. Stay down and stay put.

Remaining passive is a choice. Do not confuse passive action with surrender. Your job is to stay alive long enough to find an opportunity to either remove yourself and your loved ones from the situation, or to destroy the terrorists where they stand. Timing is critical, and a passive approach may just buy you the time you need to live.

CHOOSING TO FIGHT

Even though being passive has its merits, the passive approach is not always appropriate. There may be situations where it becomes obvious to you that there is no benefit in remaining passive and that your only chance for survival is to fight the terrorists head-on. In some circumstances you may choose to fight even though you are unlikely to survive. I can imagine that if I were to have knowledge that the terrorist in front of me has plans to hurt my family, I will choose to fight. If he wants to be a martyr I will be happy to send him on his way, even if he takes me with him.

Choosing to fight brings with it a significant chance of failure. Even if you succeed in killing the terrorist, you may lose your life in the aftermath. Such was the situation aboard Flight 93 on September 11, 2001. The choice to fight the terrorists was a brave and correct decision, for all aboard that aircraft were doomed if the terrorists carried out their plan. The fight, however, did provide the passengers with a slim opportunity to survive. Unfortunately, the passengers of Flight 93 lost their lives but in doing so they saved countless others who would have otherwise perished. The passengers of Flight 93 are heroes to all true Americans.

When you have made the choice to fight you should check around you for anyone who seems willing to join the conflict. If others seem willing to join, you should try to coordinate the attack to the best of your ability. Make sure you don't give away your intention to attack when looking for accomplices or making plans. A detailed battle plan will not be possible in most situations. Detailed plans take time, and time is something that is not on your side. Sometimes it is sufficient to exchange glances or a simple nod of the head will confirm that you are not alone in the battle. When you decide to make your move, don't wait for the others to follow. Make your move when you see the opportunity.

Timing is important in a battle. One of the most difficult problems you will be faced with is exactly *when* to attack. I want you to know that it is absolutely impossible to know the perfect moment of attack. You can only know when the perfect moment *was* after the moment has passed. Conditions change second-by-second and you very well may launch your attack just as the odds go heavily against you. Conversely, the odds can swing in your favor as well and you should make the most of the moment when this happens. Attack the terrorists when you have fully committed yourself and the opportunity arises. There will be no turning back.

When the moment of attack comes you must execute your plan with a vengeance. The success of your attack will depend upon your timing, strategy, execution, and the unpredictable. You have control of the first three of these, and the unpredictable events must just fall where they may. Time your attack when the terrorists are at their weakest point from the perspective of your strategy. The weakness may be a decrease in the number of terrorists present at a certain time. The weakness may be in the physical placement or position of the terrorists as they move about. The weakness may be a lack of attention or fatigue. Whatever strengths the terrorists possess, there will be weaknesses as well. Search out the weaknesses and use them to your advantage.

Your strategy should be clear in your mind. Others may or may not be clued in on your strategy. Either way you need to create a strategy that does not depend upon everything going perfectly in your favor. This won't happen. Create a strategy that is capable of absorbing some loss and is also capable of taking advantage of unpredictable events that fall in your favor. You won't have a lot of time to form your strategy in most cases, so do the best you can with the time you have available. All you can do is create the best plan possible given the information that is available at the moment of decision. No more, no less, no regrets.

If your plan calls for stealth, then be stealthy with drive and determination. If your plan calls for an open attack, then launch your attack explosively. Whatever you do, you must do it with complete dedication to seeing it through. Half-measures will certainly get you killed, and probably many others along with you. Your life and the mission you hope to accomplish directly depend upon how well you execute your plan. Execute your plan with a fanatical determination that will exceed the fanaticism of the terrorists you face.

COVER AND CONCEALMENT

The nature of a direct attack implies that force is being applied with the intent to harm you and others around you. In some terrorist attacks the force applied may be in the form of explosives or gunfire. If this is the case, it will be necessary for you to remove yourself to a place where the shrapnel or bullets are not likely to reach you. You may also choose to hide yourself to avoid detection. Being detected is a sure way to attract bullets and shrapnel. Cover and concealment are terms I will use in the following sections that describe ways of dealing with bullets and unwanted terrorist attention. It is important to know the difference between these two terms, and when to use one or the other.

COVER

Cover is the term I will use when I speak of something that you can hide behind or take shelter in that would be effective in stopping a bullet or shrapnel before it reaches you. Taking cover behind a large concrete pillar, for example, would provide protection from bullets and shrapnel. Brick walls, large trees, mounds of dirt, steel beams, and rocks provide good cover. Using an open car door as cover, however, would probably not be effective in stopping even a bullet fired from a handgun and would be almost useless against bullets fired from a high-powered rifle. I don't consider car doors to be cover. Even so, something is better than nothing. Use anything available to put materials between you and the direction from which the projectiles are coming.

CONCEALMENT

Concealment is the term I will use when I speak of hiding yourself from the sight of a terrorist. A concealed position may or may not provide cover. Concealment is cover only when the item you are hiding in or behind can actually stop bullets or shrapnel. Hiding beneath a table or desk in your office will provide some concealment, but not cover, because a table or desk doesn't stop bullets. Bulletproof glass is cover, but it is not concealment since the terrorist can see you perfectly through the glass. Concealment is best used when the terrorist is not aware that you are there. If he doesn't know you are there, then he won't be looking for you. Keep it that way by concealing yourself and remaining very quiet. When being chased by a terrorist you will use concealment only when you are sure that he has temporarily lost sight of you. A good opportunity is when you have turned a corner, or have otherwise momentarily broken contact. If the terrorist is right on your heels, don't even try to conceal yourself. It will be a waste of time and your energy would be better spent putting some distance between you and the terrorist, finding a weapon, or continuing to flee until a better opportunity comes along. When the chance to conceal yourself does come, you must immediately duck into your concealed position and get still and quiet quickly.

As you go about your day, it would be useful to take a look around and identify objects that provide cover or concealment or both. Learn to find these areas of cover and concealment in a hurry, and to know the difference between the two. Effective use of cover and concealment can save your life.

WEAPONS

At this point I will turn my attention to some of the weapons used by terrorists. The following discussion of weapons is intended to provide you with a general knowledge of the items that may be used against you in a direct terrorist attack. Consequently, these same weapons can be effectively used against a terrorist. The sword cuts both ways and the more you know about the capabilities and limitations of these weapons, the more effectively you can use them and defend against them.

You've seen it over and over. A television actor engages an assailant. There are a lot of impossible moves, lots of talking, and a lot of grandstanding. Eventually the terrorist loses his weapon and the "good guy" throws down his own weapon just to make the fight "fair". Bull! I can tell you with certainty that I am not interested in the television version of this battle. If the terrorist loses his weapon and still poses a threat, I will kill him with my own weapon and then take his weapon. If his AK-47 falls into my hands I will empty its magazine into every terrorist in sight. If a terrorist dies in battle, I will pick up his weapon and continue the fight to annihilate his buddies. Never leave a useful weapon behind. It could be picked up by your enemy and used against you.

As you will see, I will alternately move from defending against these weapons to using them during a terrorist attack. My reason is clear. I expect you to use any weapon available to your advantage, and this includes the weapons of the terrorists themselves.

Chemical Sprays

Chemical sprays are widely used as non-lethal self-defense weapons. Most of these chemical sprays in America today contain *oleoresin capsaicin* (OC) as an active ingredient and are commonly and collectively known as pepper sprays. Other sprays may contain CS or CN agents and are known as tear gas. Both sprays have strengths and weaknesses.

Tear gasses containing CS or CN agents act on the principle of being a painful irritant. The drawback to this principle is that a person able to withstand pain or under the influence of a pain-inhibiting substance may not be impaired significantly. Pepper sprays are inflammatory agents that cause swelling in the mucous membranes, nose, and throat. Pepper sprays will likely shut down all breathing except that amount necessary for life support. The drawback with pepper sprays is that the spray must hit its target fairly square in the face to be effective. Some sprays contain a mixture of the two agents (OC and CS or CN) and these are probably the most reliable on the market.

Pepper sprays and tear gasses cause a severe burning sensation on the skin, mucous membranes, and eyes and may impair breathing. They will not kill you, but you will be very uncomfortable if you are sprayed with the agent. It is important to remember that these sprays are non-lethal agents and that you can actually continue to fight even though you may be impaired.

To make my point I will relate to you an experiment to test the effectiveness of pepper spray on my ability to continue to function. I can normally fire eight rounds in ten seconds from my Ruger .45 caliber at a target 25 feet away and place each round in the center ring. At a local firing range where I was the only shooter I had a friend spray me in the face with pepper spray. I gave it two seconds to take effect (and it certainly did) then turned to my target. I drew my weapon, fired eight rounds in fifteen seconds and placed three of the eight rounds in the center ring, two in the next ring out of center, and the other three in the second ring out from the center ring. Each shot, however, would almost certainly have been fatal to any terrorist not wearing body armor. My point is this: even impaired by the pepper spray I was able to function in a manner sufficient to cause severe harm to a terrorist. Yes, it was uncomfortable and extremely painful. Yes, it reduced my speed and accuracy with my weapon. No, it did not stop me from doing what I needed to do.

Knowing this, I am always hesitant to suggest the use of chemical sprays as a serious form of self-defense. I do believe that in many cases they are useful defense tools when used properly, and they are probably better than nothing. On the other hand I know that chemical sprays are not always capable of stopping an attacker. This is especially true if the attacker is very committed, drunk, or on drugs. Chemical sprays are also susceptible to wind currents. Be mindful of this fact to avoid disabling yourself with your own spray. Police officers use pepper sprays and tear gasses, but they also have a baton, handgun and other devices as backup. If you choose chemical sprays as a form of self-defense I would suggest that you speak with a local police officer to get his or her opinion as to which brand you should use and a few tips on how to use it.

Knives and Clubs

Knives and clubs do not at first glance seem to be very effective weapons in committing the mass of death and destruction that would collect headlines for a terrorist's cause. After all, a knife or club is not exactly a weapon of mass destruction, or so it was believed before September 11th, 2001. The weapon of choice was the extremely sharp, very compact box cutter. With this simple device the terrorists slashed passengers and gained access to the cockpits of four jetliners that became guided missiles against our fellow Americans.

If you are faced with a knife or club, the best bet is to stay out of reach of the weapon. Knives and clubs have a very limited effective range and you'll do well to stay beyond that range. Use anything within reach to block the blows of these weapons. If you become cornered in such a way that you cannot remain beyond the range of the weapon then your tactics must change, depending upon whether you are being attacked by a knife or a club.

If you are faced with a club, the terrorist must have enough room to swing the weapon for it to be effective. If you can't stay out of the terrorist's reach, then your best bet may be to charge into him and stay as close and as high on his body as possible. Grabbing the terrorist by the waist, knees or ankles won't help unless you can bring him down immediately. By grabbing too low you will open your back, head and arms to very powerful strikes from the club. Grab and hold high around the chest, shoulders or neck. Grab him in a bear hug and don't let go. He may hit you in the back of the legs with the club, but these injuries are not life threatening. While you're hugging, you'll want to push against anything you can reach with your feet. If there's nothing to push against then kick him with your feet and jam him with your knees with all the force you can muster. Be violent! Scream, kick, head-butt, bite, choke and claw until you have an opportunity to break your contact with him cleanly. When that opportunity comes, move out of range of the club immediately and put as much distance between you and the terrorist as you possibly can. If you have determined that you cannot take a chance on releasing or restraining the terrorist, go for a choke or an eye-gouge and maintain your contact. Use any weapon available (including his) without hesitation and don't stop until you are sure he cannot continue to be a threat...or until he is dead.

Defense against knives will require a different tactic. Whereas a club's power is reduced if you are too close for the terrorist to get a good swing, a knife will be lethal at very close range. If you are not trained in knife defense tactics, do not attempt to grab and hold a knife-wielding terrorist unless you are trying to buy some time for others to escape. Be aware, however, that holding onto a terrorist who is using a knife will likely end your life. If you become cornered and have no weapon, attempt to stave off the attack by kicking your attacker. Be sure to make your kicks quick and low and do not allow the terrorist to grab your foot or leg; that could bring horrible consequences. If you do become entangled with the terrorist, try to keep him as defensive as possible by raining blows into his face and eyes. Try to get enough room to break your contact with him. Accept the fact that you will be cut or stabbed (or both) if you come into contact with a knife-wielding terrorist. You will see blood, and it may be yours. Don't stop and don't give up. Many superficial cuts bleed like crazy, but they are not life threatening. Just because there's blood doesn't mean that you've lost the battle. It simply means that you're still alive! Keep fighting and keep moving. Something will break loose eventually and you can make your escape or put an end to the terrorist.

If given the choice between a knife and club when fighting a knife or club, I would prefer to use a club. A club makes an effective blocking instrument and can be useful from a little greater distance than a knife. The club, too, can deliver powerful shocking blows to any part of the body that comes into its range. Concentrate blows with the club on the upper body, preferably the shoulders, neck, head and face. This will keep the terrorist defensive. Be quick and decisive with your blows and never allow the terrorist to take hold of the club. If he grabs the club you should immediately kick him low and attempt to wrest the club from him. Keep a firm grip on your club at all times and don't sling it away by letting it slip from your hand during a strike. When striking with your club, be sure to go for specific targets with all of the fury you have in you. Select a target…the nose, the head, the cheekbone…and drive your weapon into your target with all you've got. If the terrorist throws up his arm in defense, break his arm with your blow and continue to strike again and again. If the terrorist turns away, level hard blows to the back of the head, base of the skull, neck, and even attack his spinal column. When attacking the spinal column, attack the bony, exposed middle and lower part of the spine. Don't waste too much time on the upper back and shoulder blades. They are fairly well protected areas. Keep your blows fast, hard, and relentless. This is not a game. He will kill you if he has the chance. Don't give him the chance.

If you have a knife, and you are fighting a terrorist armed with a club, it's best to get in very close. Get too close for his club to make effective contact. Use your knife to slice and stab at any part of his body that comes available. His attack on you will end quickly if your knife finds his throat, or is stabbed deep enough to puncture or sever the heart, aorta, or femoral arteries (on the inside of either thigh). A cut to any of these vital areas can ultimately be fatal to the terrorist. Even so, never let down your guard until you are absolutely certain that the threat has been removed.

When fighting using a knife it is best to keep your knife close to your body and in a firm grip. Strike with your blade and then retrieve your weapon quickly. Don't hold the knife out in front of you as is so often seen on TV; this will lead to a kick or blow that will cause you to lose your weapon. You must protect the weapon to keep the weapon, and you must keep the weapon to protect yourself.

When you are combating a terrorist up-close and personal, also known as hand-to-hand combat, concentrate your energy on vulnerable targets. It is a natural response to protect the face and especially the eyes. Use this natural, reflexive response to your advantage. Fire rapid, hard blows with your fists into the eyes and face of the terrorist to attempt to make him defensive. If you don't have room to draw a punch, dig your way with your fingers to his eyes. This is no time to be squeamish. Blind him if you can.

When you are combat committed there is no negotiation or backing out. Once the decision has been made to engage in physical combat with a terrorist there is no good way out except to win. This is not a street fight where you can "prove your point" and then walk away. This engagement will likely be the end of one of you. Make sure it isn't you by getting this thing over with as quickly as possible by using all of the force and determination you can gather. Indeed, your life does depend upon it.

Firearms

Firearms in the United States have been a sharp point of controversy over the years. I will not get into the Second Amendment arguments here because those arguments are beyond the scope of this book. Terrorists do not care about the legal and political arguments. They know firearms are effective tools for both offense and defense and they will continue to use them so long as they can get them. The supply of firearms on the world market is endless for terrorists. Police departments and the military also know the usefulness of the firearm. Otherwise, the firearm would not be standard issue.

If you do decide to purchase a firearm please have the presence of mind to seek high-quality training from qualified personnel. You owe it to yourself and everyone around you to be proficient with your firearm. Concentrate on real world shooting under different conditions that you might encounter. Learn to hit your target while holding your weapon with one hand, both hands, either hand, in dim lighting, with distractions, while targets are moving, etc. Learn to load and reload under stress, with one hand, in the dark, quickly. There are schools that teach these techniques. Attend them. Learn from them. Practice regularly. You owe it to everyone around you. The following discussion is no substitute for the training you must seek if you own and carry a gun.

Now we need a little firearm refresher. Handguns are firearms supported and fired by using one or both hands as a sole means of support. Rifles are firearms supported by both hands and one shoulder. Handguns are relatively compact as compared to the much longer rifle. Handguns, by virtue of their size, generally fire less powerful ammunition than do rifles. Some ammunition, however, is suitable for either handgun or rifle. The effective range and accuracy of a handgun is less than that of a rifle. Shotguns are shoulder-fired firearms like a rifle, but they differ in the fact that the ammunition releases several pellets instead of one projectile. The effective range of a shotgun is less than that of a rifle, but the shotgun covers more area with each round fired (because the pellets spread out after leaving the barrel).

Handguns, rifles and shotguns come with various mechanisms to feed the ammunition into the firearm. Generally, they are divided into automatic, semi-automatic, and manual. There are many variations on each of these types of fire-arms, and I won't get into all of the differences here. A fully automatic firearm allows you to press the trigger down and it will continue to fire and reload for as long as you hold the trigger down or it runs out of ammunition. This type of fire-arm is sometimes called a machine gun. A machine gun can fire several hundred to several thousand rounds per minute. A semi-automatic firearm fires one round each time you pull the trigger, automatically loading the next round of ammunition in preparation for the next pull of the trigger. If you want to fire five rounds from a semi-automatic rifle or handgun you will have to pull the trigger five times. A manual firearm will fire one round of ammunition each time the trigger is pulled, but it does not necessarily load the next round of ammunition for you. Depending on the mechanism, another round of ammunition may be placed in position by a revolving cylinder, the crank of a lever, the sliding of a bolt, or by hand.

Most of the military rifles that are used are of the automatic and semi-auto-matic variety. Military shoulder-fired rifles like the M-16 and the AK-47 are commonly called assault rifles. It is important to note that assault rifles are no more powerful than other types of rifles. In many cases the assault rifle will fire less powerful ammunition so that it may be controlled when fired in fully auto-matic mode. For example, the M-16 uses .223 caliber ammunition. This is a high-velocity, small-bore cartridge that does not pack near the punch of a com-mon big-game rifle like the 300 Magnum. On the other hand a 300 Magnum is not likely to be very controllable if it were firing 600 rounds per minute. The danger presented by an assault rifle is more from the number of rounds it can rap-idly fire and the capacity of its magazine than in the power of the cartridge.

Remember this: the most powerful, fastest shooting assault rifle, handgun, or big-game rifle in the world is useless if the bullet doesn't hit its target. I have been on firing ranges all over this beautiful country and I have seen in each of them the odd person cranking out dozens of rounds per minute at a silhouette target 25 feet away. Brass and lead from the semi-automatic handgun fills the air as each round is fired. After three or four magazines of ammunition have been expended the proud shooter withdraws his target to display the "3 good hits" he made. He can hardly contain his excitement. My question is this: what happened to the other 42 rounds that went downrange? The answer? Wasted. If this had been a real emergency and this person was aiming at a terrorist, the terrorist would have been the least likely person to be hit. Don't be this shooter. If you don't think you could ever shoot a terrorist in self-defense, or if you do not intend to be expert with your weapon, don't carry one at all. If you're going to carry a weapon, do us all a favor and learn to use it, be proficient, and be on target first time every time. One well-placed round is much more effective and safer than a hundred that miss.

This brings me to my next point: if you come under attack by a terrorist with a firearm, make him miss. The keys to making a terrorist miss his target are movement, distance, and cover. As you can imagine, it is much harder to hit a moving target than a still target. If you come under fire, or you believe that an attack with a firearm is imminent, get moving. If you aren't in a position to get behind cover, then you should be running, dodging, and zigzagging toward cover. Your direction of flight should be away from the terrorist if at all possible. It is a fact that the greater the distance, the harder a target is to hit. Make that distance as far as possible as quickly as you can. This is especially true when it comes to handguns. A handgun's short barrel makes it highly susceptible to small changes in the aiming point. A small sighting error at the muzzle of a handgun translates to a large error at a distance. A bullet that is off-target by 3 inches will still strike you with deadly consequences if you are only 3 feet from the muzzle of the gun. That same bullet will be off-target by nearly 2 feet when it is 20 feet from the muzzle of the gun. At that range the same bullet would probably miss you completely. The average person can run 20 feet in less than 2 seconds. Use this to your advantage by getting as far away from the firearm as possible and taking cover as the opportunity arises.

It is important to note that the trajectory of a bullet is a straight line horizontally, and a slight curve vertically. The vertical curve of the bullet's trajectory is so slight that it can be ignored when trying to escape direct fire from a firearm. This means that anything you can do to put something between you and the armed terrorist is very helpful. He can't shoot around corners or over walls. Turn corners, weave in-and-out of columns, or put anything between you and the terrorist as you flee. If you dive to the ground or happen to fall as you flee you should roll to the right or left as soon as you make contact with the ground. The terrorist will likely be raining bullets down on the point where he saw you fall. Make sure you aren't there when the bullets arrive.

In your flight you can also make use of concealment. Concealment is not protection from the bullet, but rather concealment makes it more difficult to hit you because the terrorist can't see exactly where you are. If you are concealed and the terrorist doesn't know where you are, it is possible that he won't even shoot at you. Keep your position for as long as it is beneficial, but no longer. If you are detected and the bullets start flying you need to get moving again. Concealment is good, but cover is better.

If you happen upon a firearm, or have been fortunate enough to take the firearm away from a terrorist, use it. The Hollywood notion that you would pick up a firearm and toss it out of reach of the terrorist is foolish. Having the weapon in your hand and emptying the magazine into the terrorist is much more effective. Take hold of the firearm, point the front sight at the terrorist, and pull the trigger! If it fires, good! If it doesn't, nothing is lost. Even if the weapon malfunctions or is out of ammunition it is a pretty good bludgeoning tool. Get busy.

I want to warn you about a complicated situation that can arise when using a firearm against a terrorist. If you are combating a terrorist with a weapon of any kind, but especially a firearm, and liberating forces or police arrive on the scene you could be in a real sticky situation. Liberating forces and police officers will probably consider anyone with a weapon in their hands to be a terrorist. If friendly forces tell you to drop your weapon then drop it, yell that you are not a terrorist, and get down on the ground. You may already be under fire from the terrorists, don't compound your situation by coming under fire by friendly forces. If you believe that you can't drop your weapon for some reason, then that is your call. I just want you to be aware of the possible consequences.

It should be clear by now that you can indeed survive a firearm attack. You will be the judge of what actions you should take to save your life. Sometimes talking it out will work well, other times it won't. Each situation is different and you must remember that you will make the best decision you can with the information available at the time. Make your decision and execute your plan with dedication. Above all, use your brain.

Improvised Weapons

An improvised weapon is any instrument found in your environment that can be used to repel or initiate an assault that is not normally employed in that manner. Improvised weapons are prolific in your environment. Learn to recognize them.

A word of caution: improvised weapons as seen on television are, for the most part, fanciful creations by persons who have never experienced hand-to-hand conflict. Don't believe the foolishness you see on television. If you have a beer bottle, don't even try to break it on a table and wield it as a sharp-edged weapon. Chances are the bottle won't break. If it does break, it'll probably cut your hand to the bone. The flying shards may well end up in your eyes and blind you. If you insist on breaking the bottle, break it by bludgeoning the terrorist over the head with it. An unbroken bottle is much more effective than a broken one. Keys held between the fingers are another favorite of television. Keys protruding between the fingers look menacing, but they are not anchored in any way in the palm of your hand. Punching with keys held in this manner will likely result in damaged and bleeding fingers. Hollywood stunts only work on the soundstage. Don't bring these stunts into the real world.

When you choose a weapon, make sure it truly is a weapon and not just a pacifier. What I mean by this is that a true weapon is one that is really capable of doing significant damage. A pacifying weapon is one that makes you "feel" better and protected, but in reality is useless for defense or attack. Worse yet, the pacifying weapon is likely to give you a false sense of security. This false sense of security can lead you to act in a way that is likely to prove fatal for you when you confront a terrorist. You can be sure that the terrorist will know whether you are holding a weapon capable of inflicting damage or not. Make sure that you, too, know the difference.

Now for a discussion on real improvised weapons. I prefer bludgeoning instruments to nearly anything else in an improvised weapon scenario. A bludgeoning instrument is one that is capable of crushing most anything in its path. A bludgeoning weapon should be heavy enough and hard enough to cause damage, but not so heavy that you are unable to swing it swiftly and accurately. Ideally it will be of a shape that allows you to hold it securely with one or both hands. Whiskey bottles are good in this way. They are heavy and the neck of the bottle provides a secure handle. Table legs, chair legs, tire tools, pipes, and nearly anything hard, handy, and moderately heavy are good candidates for use as improvised bludgeoning instruments. Though not heavy, a car antenna can be broken from its mount and used in a whipping manner. Striking a terrorist in the face with a car antenna can cause a gaping laceration. Be careful with this one, though, it probably won't stop him cold.

Improvised weapons also include edged weapons and pointed instruments. Pencils, pens, blades, and any sharp-edged item you can easily wield are good candidates for this category of improvised weapons. I have a word of caution concerning pointed and bladed instruments. You will probably have to be in very close proximity to the terrorist to use these items effectively, and the effects of the instrument on the terrorist may be delayed. Unless you cut or stab into a vital blood or nerve pathway, the terrorist may still be able to function long enough to bring a catastrophe down upon you. Even a fatally stabbed terrorist may not die for several minutes. During much of that time he may be capable of doing harm to you or those around you. Be aware of this situation and do whatever you must to ensure your safety.

Thrown objects comprise an additional category of improvised weapons. I don't usually encourage throwing weapons at terrorists since I'd rather have the weapon in my hand. Even so, a pot of scalding water or coffee (like you may find in an aircraft kitchen area) could be a temporary deterrent. You might throw cosmetic or baby powder, dirt, or liquid into a terrorist's eyes, or aim whatever is at hand in his direction. Concentrate your aim on the face and eyes and throw whatever you have with all of the force you can muster. Aiming at the face will force him to be more defensive than if you are throwing anywhere else on his body. You'll probably even want to aim a bit low since he'll attempt to duck under your projectile. The most important thing to remember about a thrown object is that it is only a temporary solution. Most of the time your object will have only minimal influence on the actions of the terrorist, and it will certainly attract a response. It is for this reason that you must not base your entire defense or attack on thrown objects. Throwing an object may be a way to open up the attack, or as an available defense, but it is not an "end-game" tactic. You will certainly have to follow up with a more substantial attack or immediately flee from the terrorist, whichever is in the interest of your mission.

Surviving Explosives Attacks

As described previously, explosives attacks are those in which devices are detonated with the intent of causing casualties with the initial blast and shrapnel, or by the failure of a building or structure damaged by the blast. The intent is to create casualties by the explosion and shrapnel itself, or to cause the collapse of a structure containing potential victims.

CONVENTIONAL EXPLOSIONS

Conventional explosions occur when extremely combustible chemical substances are confined and detonated. Conventional explosives can be created with little technical expertise. The more powerful and compact the explosive, the more expertise required. Conventional explosive devices may injure or kill their victims through the initial shock wave, but most casualties will occur because of shrapnel. Shrapnel is any object added to an explosive device that is to be fragmented and thrown by the force of the explosion. Shrapnel is normally propelled at a high velocity. Shrapnel may be part of the bomb's casing, or a bomb may be placed inside another container that includes objects that become shrapnel. A grenade is an example of shrapnel that is included in the bomb's casing. Suicide bombers, on the other hand, often pack explosives inside vests that are worn on the body. The vests are filled with bolts, drill bits, steel ball bearings, or other metal and glass objects that become shrapnel when the device is detonated.

It would be highly unusual for you to receive more than a few seconds' warning before a conventional explosion is detonated. You may get no warning at all. A scenario that has played out many times the world over has been the use of hand grenades against civilians. Mass transit vehicles, crowds, and even places of worship have been subjected to grenade attacks with alarming frequency. This situation may present you with only a couple of seconds of warning.

The typical, unmodified grenade will have a fuse that detonates the grenade approximately four seconds after the spoon is released. The spoon is the handle that runs along the side of the grenade body. The spoon is released as the grenade is thrown, which arms the grenade. The grenade is not armed by pulling the pin and ring assembly, but rather by the release of the spoon. That's why a person can hold a grenade for an indefinite amount of time after the pin is pulled. The grenade will not begin the (approximate) four-second countdown until the spoon is released. Properly thrown, the grenade will bounce around so that it cannot be easily caught and thrown back at the attacker. Never try to grab a grenade. At most you'll have a second or two, which is not enough time to play catch and throw. A second or two, however, is enough time to put your body into a better position to survive the blast.

If you are unlucky enough to be the target of a grenade attack, but lucky enough to have a couple of seconds before it explodes, use this time wisely. You must respond immediately by diving to the ground in the opposite direction of the grenade. Your head should be away from the grenade and your feet should be toward the grenade and held together. Clasp your hands around your head and the back of your neck. If cover is available and convenient, then make your dive behind cover. Don't try to make it to cover thirty feet away. The distance is too great and your time is better spent just getting to the ground with your feet together and toward the grenade. When the blast comes any shrapnel that does hit you will be more likely to strike the soles of your shoes (probably the most protected part of your body). The point is that your feet and legs are shielding more vital parts of your body from the blast and shrapnel.

The response for grenade attacks is applicable to most other explosives attacks as well. If you are in a position to see that a suicide bomber is about to trigger an explosion and you believe you have only a couple of seconds, then your response should be exactly the same as if you saw a grenade being thrown. Understand, however, that a suicide bomber will be carrying a much more powerful explosive device than a grenade. You may get no warning at all. In this instance your response will likely be to provide emergency first aid to yourself or those around you. Unfortunately in some situations you can only respond after the event.

There have been occasions where a suicide bomber is caught in the act of attempting to trigger the explosive device. Such was the case with Richard Reid on a flight from Paris to Miami, and with a would-be suicide bomber attempting to board a bus in Tel Aviv, Israel. In both cases the suicide bomber was attacked and subdued without detonating the device. In the attempted bus boarding, however, the bomber managed to get away and run to a bus stop where he detonated the explosive, killing one person in addition to himself. In the case of Richard Reid, diving for the deck of the aircraft would not help since the aircraft would have been blown from the sky. There have been other instances where the bomber was able to detonate his explosive device even though he had been tack-led. This is a risk, but it may be your only chance to survive. The only guarantee is that by taking no action you will certainly become a victim.

The cruel fact is that if you are close enough to tackle a suicide bomber, then diving for cover may be of limited use anyway. Being that close to the detonation of twenty pounds of plastic explosive is almost certain to be fatal. Your best hope of surviving is to tackle him in as violent a manner as possible in an attempt to keep him from detonating the weapon. Get hold of his hands at all cost, and don't let him pull any cords or press any buttons. Pin him down and don't let go. He has already committed himself to death. Don't let him take you with him.

I know you have all seen the movies that show a soldier diving onto a grenade or other explosive device to save others. This act has heroic and well-placed inten-tions. In reality, it may reduce the amount of force that is distributed in all direc-tions and, in turn, save some lives or injury. The soldier, of course, will be killed instantly. To maximize the protection of an individual you are committed to sav-ing (such as a child or spouse) it would be more effective to dive onto the upper body of the person you are protecting instead of diving onto the grenade. This is especially true if the grenade is more than a couple of meters away since you wouldn't have time to reach the grenade anyway. To protect a loved one, you should wrap yourself around their head and shoulders with your back to the device. You then become a shield against the shrapnel. Don't hug the person in the normal head-to-head fashion because their head will be left unprotected over your shoulder. It's better to give the person's head a "bear hug" and be sure to place yourself between the person you're protecting and the coming explosion. All of this must be done in a second, of course. Don't worry about grabbing too hard. You should be aware that it is likely that you will be seriously injured or killed by doing this, but it may give your loved one a chance to survive.

Some establishments frequented by American patrons have become the target of terrorists using car bombs, bicycle bombs, and package bombs. The terrorists have recently added a deadly twist to these public bombings. The terrorists have staged the bombs into two separate events. A smaller bomb is detonated to create a few casualties, which brings many people and emergency crews to the scene. As the crowd gathers to help the victims of the first bomb a larger, more powerful bomb is detonated. This has a psychological effect as well as creating more casual-ties. Be careful around the scene of a bombing and be aware of this tactic. Unless you are actually participating in the rescue efforts or you are injured and immo-bile, leave the area as quickly as possible.

Conventional explosives in the hands of terrorists have proven deadly on an almost daily basis. The best way to survive terror is to recognize terrorist action and stop or avoid it before it starts. By the time the grenade is thrown or the suicide bomber has boarded the bus it is probably too late. You must be vigilant and take whatever action is necessary to either stop the bomber yourself, or alert others of the threat.

NUCLEAR EXPLOSIONS

Nuclear explosions occur when a nuclear device is detonated. The possibility of nuclear weapons falling into the hands of terrorists has become a major concern in the past few years. The technical expertise required to build, maintain, deliver, and detonate a nuclear weapon is considerable and costly. Even rogue states with seemingly endless resources have had difficulty bringing their nuclear weapons dreams to life. Even so, it is not beyond the realm of possibility that a terrorist organization might gain access to a nuclear weapon through a variety of means. The thought that terrorists might obtain and deliver a weapon of this power is a grave and real concern.

The nature of nuclear weapon technology is such that the smaller a nuclear device is, the more difficult it is to create. Making a nuclear weapon the size of a railcar has fewer technological hurdles to overcome than creating a nuclear weapon the size of a suitcase. It is for this reason that it may seem unlikely that a compact, high-yield nuclear device would be detonated on United States soil. On the other hand, consider the size of the containers on a cargo ship. They are the size of railcars. A ship sitting in the Port of Oakland, Long Beach, or New York awaiting inspection could carry a nuclear device in its hold. The results of a nuclear detonation in this circumstance would be catastrophic. There really is no need for a terrorist organization to create a suitcase-sized bomb when the lower-tech solution would work just as well.

The first product of a nuclear detonation is light. If the explosion happens at ground level or above, the flash of light may be seen for miles. This flash will travel faster than any other product of the detonation and will arrive before any of the other damaging effects of the explosion. The amount of time there is between the initial flash of light and the heat and shock wave that will follow is dependent upon how far you are from the point of detonation. The farther you are from the explosion, the more time you will have between the flash of light and the heat and shock waves. In any case, the amount of time will be measured in seconds so you must take cover immediately.

The second product of a nuclear detonation is heat. Extreme heat is produced at the point of detonation and it will be radiated in all directions emanating from the nuclear fireball. The temperature of the fireball will be measured in millions of degrees and the heat wave will be capable of igniting and incinerating materials for great distances. The greater the distance from the fireball, the less extreme the heat will be. Any combustible materials within a mile of an even moderately sized nuclear detonation could burst into flames. Persons who are not incinerated immediately are likely to suffer the most severe burns imaginable.

The third product of a nuclear detonation is the shock wave. The shock wave is a highly compressed wall of air moving at an extreme speed. Close to the explosion the shock wave will likely destroy nearly everything in its path. Farther from the center the shock wave will be carrying shards of material picked up along the way. This is one reason you should take cover immediately. The material carried by the shock wave is capable of ripping you apart. Be aware that the shock wave will first travel out from the explosion and then reverse course as the vacuum created by the explosion begins to fill. So a severe shock wave will cross over your position coming from the direction of the explosion, then a less severe but still powerful shock wave will cross your position again traveling toward the explosion.

The fourth product of a nuclear detonation is a massive amount of radiation. Initial radiation occurs within the first minute or so of detonation. Initial radiation is extremely intense. Anyone in the immediate area of the detonation (who happens to survive the heat and shock wave) is likely to receive a debilitating or lethal dose of initial radiation. In a nuclear explosion large amounts of soil and water become vaporized and the particles become radioactive. These particles, and any remnants of the explosive device itself, become suspended in the air where they are carried by the wind. The particles that have become radioactive are collectively known as "fallout". Fallout is a radiological agent and is extremely dangerous. Fallout begins to settle within twenty minutes of detonation and may continue for hours or days.

At this point I want to address your response if you happen to be unfortunate enough to witness that flash of nuclear light. You should immediately make a dive for the nearest substantial cover while getting down close to the ground. Substantial cover can be a ditch, a concrete pillar, the base of a building, or beneath a table or automobile. Don't waste time making a decision. If you don't see cover immediately, then just hit the ground where you are. Don't try to make it to a parking garage fifty feet away. You won't make it. Take whatever cover is available within a few feet of you. Cover your face and head with your hands and arms and hang on. This is the old "duck and cover" wisdom of the Cold War. Stay in your covered position and don't get up until you are confident that the heat and shock waves have passed. Don't look at the fireball directly. The intense light could blind you. When you do get up, you should immediately move to limit your exposure to radiological fallout. I discuss some things you can do to reduce your exposure to fallout in the chapter titled *Surviving Distributed Agent Attacks*.

STRUCTURAL COLLAPSE

Terrorists have often used explosives attacks to cause the structural failure of buildings impacted by the blast. In addition to the collapse of the structure, fire may spread rapidly as gas escaping from broken pipes and flammable debris in the rubble ignite. The best plan is the plan you have made *before* the attack happened. Know where the emergency exits, fire extinguishers, and first aid kits are before disaster strikes.

If a building you are in is shaken in such a way that you believe the building could collapse, you should duck under anything substantial enough to provide protection from falling debris. Heavy desks and tables are the most common items that may provide some protection in office buildings. If nothing is available you should move quickly toward a wall or doorway and squat against it, covering your head and neck with your hands and arms.

In the event that the building (or part of it) does collapse there will likely be a lot of dust in the air. The dust from a building collapse is not the same kind of dust you will find along a country road. The dust that emanates from structural failures contains ground glass, concrete, and other particles that could lacerate your eyes and lungs. Don't breathe this dust directly. Instead, cover your mouth and nose with your clothing or any available cloth. Try to filter the dust through the cloth as much as possible. You'll need your eyes if you are trying to move about, so protect them as much as you can. If you are not moving, close your eyes until the dust settles a bit.

If you have become trapped you should not attempt to yell for help unless you are certain someone who can help is very close-by. Yelling will only force you to take in large amounts of the dangerous dust that is filling the air. You could be trapped for a long time. Don't make things worse upon yourself by filling your lungs with sharp and toxic particles. To attract attention you should tap any object you have against the wall or a support beam in a steady, slow rhythm. Search crews will be listening for this sound with sophisticated instruments as an indication that someone is trapped. Use this fact to your advantage. Save your breath, save your strength, and continue tapping.

If you have become trapped in the building but are still able to make it to a window you may be in luck. Hang a cloth out of the window to signal rescue crews. Bright clothing works well for this purpose. If the phone is still working, call 911 to inform the rescuers of your position in the building.

If you are trapped and the building is on fire you should close any doors that you can. Use whatever is available to seal vents and doors. Cram towels or clothing into large cracks and tape smaller cracks. Smoke is a deadly enemy; keep it out. Close or open windows as the situation mandates. Remember, the fire is constantly changing. The window may need to be open one moment, but may need to be shut as the situation changes. Don't break windows if at all possible. You may need to seal them again later. Often smoke will begin to fill an area long before the fire arrives. Smoke is a killer just like fire. You should get as far away from it as possible. Smoke rises, so you should stay close to the floor where the cleaner, cooler air is. Breathe through your clothing or a cloth and crawl to the exits or windows. An intense fire could create enough heat to cause any moisture in your cloth to become steam. Breathing steam can damage your lungs. A dry cloth is safer than a damp cloth in an inferno.

When you approach a door in a burning building you must test the door for heat before opening it. To test for heat on a door you should touch the back of your hand to the door. Don't ever use your palm or fingers to test for a hot door. If the door is very hot you could be burned severely. It is better to burn the back of your hand than the palm and fingers since you'll be needing the palm and fingers to help you grasp and manipulate objects to save your life. If the door is cool you can open it slowly. Be ready to close it quickly if you find thick smoke or fire on the other side. If it is safe to pass, then exit and close the door behind you. Close all doors behind you to help contain the smoke and fire.

If your clothing has caught fire you should remember these three words: Stop. Drop. Roll. If you are on fire you must immediately stop where you are, drop to the ground, and roll until the fire is smothered. Don't ever run when you are on fire. Running increases the available oxygen feeding the flames on you and your clothing. In essence, running simply "fans the flames" and will certainly make the fire much worse.

When attempting to move about and escape a building that has become structurally unstable or is on fire you should seek stairwells and avoid elevators. A flashlight will be a handy item to have. Beware of any broken or exposed power lines that could electrocute you. Stay away from broken steam pipes and any pipe that is expelling unknown substances. Even a small, seemingly insignificant stream of high-pressure steam can cause severe wounds. Finally, use your best judgment when making your way through the rubble. There may be unstable portions of the building that stand between you and safety. You will have to make the best decision you can with the information available when the time comes to move out or stay put. Only you can make this call, and only you can know whether the risk of staying put is greater than the risk of crossing unstable and dangerous ground.

Explosives attacks present a complex set of problems for anyone caught in the explosion or its aftermath. Your best chance at surviving an explosives attack is to be prepared to take immediate action to save your life. Sometimes that may mean diving for the deck or tackling a would-be bomber. Other times you will simply be trying to survive in the aftermath of the explosion. Stay alert, stay committed, and stay alive.

Surviving Attacks on Mass Transit Vehicles

Mass transit systems, for our purposes here, will be defined as any vehicle and the supporting infrastructure that carry large numbers of people along prescribed routes as a matter of routine. This definition suits the topic well for this chapter because I am going to be dealing with airliners, buses, and trains that may become targets of terrorism.

BUSES

Buses are a part of the public transportation system that have seen their share of suicide bombings, especially in Israel. Typically a suicide bomber will board a bus and detonate a device that was carried aboard in a duffel bag or underneath the clothing in a specialized bombing vest. The blast is usually horrific, and the shrapnel maims and kills many aboard. There may be little or no warning in this type of attack. More often than not, the victims of suicide bombings are left to respond to the cries of injured survivors if they themselves have not been seriously injured or killed in the blast.

There may be situations where you have some warning, or you have become a hostage on the bus. In either case it is helpful for you to be familiar with the escape routes and any potential weapons on the bus. First, take a look at the windows. You will usually see several windows that are designated as emergency exits. Become familiar with the operation of these windows. Which way does the handle pull? Does the whole window pop out, or does it just swing out on a hinge? Where is the nearest emergency exit? Remember to look behind you. What about the ceiling? Are there any vents up there that are designated or otherwise would serve as an escape hatch? If the bus overturns the ceiling may provide a good escape route. Where are the fire extinguishers? Fire extinguishers are good for putting out fires and cracking terrorists on the head; whichever suits the situation. Is there a fire axe or other tool on the bus? I think you get my point. Start becoming familiar with your bus the very minute you board. You may not have time to study it later.

RAIL TRANSPORTATION

Commuter trains and trams serve as a quick and efficient way to move a multitude of people through metropolitan areas. Though to a lesser extent in the United States than in Europe, trains provide long-distance travel along busy corridors. In most cities the trams alternately run above ground and underground. It is in this capacity and context that subway trams provide an attractive target for terrorists.

Consider the problems that are solved by the subway for the terrorist. In a chemical or biological attack one of the major problems faced by a terrorist is the distribution of the agent in sufficient concentrations to affect or infect the intended victims. Subways solve this problem creatively by concentrating large numbers of people in the enclosed tunnel. An agent disbursed in a subway tunnel will have thousands of passengers per hour passing through it. In the case of biological agents in powder form (such as anthrax spores) the motion of the moving train can actually facilitate the spread of the agent as the wind whips the agent farther along the tunnel and into the ventilation systems of the trains and stations.

The point is that there is a high concentration of people moving in a captive environment on a predictable schedule. These facts did not escape the Aum Shinrikyo cult in Tokyo, Japan, where twelve died and thousands were injured by sarin gas distributed in the subway. As horrible as this attack was, the outcome could have been much worse. In the future a much more devastating attack is almost a certainty. Subways and trains could be targets of explosives attacks or even the target of automatic weapons fire. The possibilities are limited only by the terrorists' imagination.

Given the obvious dangers you may face in a subway or train, it would be a good idea to familiarize yourself with some of the essential elements of the vehicle. Notice that there will usually be an "emergency stop" switch somewhere in each car of electric trams. There will be designated exits, fire extinguishers, and occasionally tools to be used in case of an emergency. Take note of where the emergency pull handles or switches are that will allow you to open the doors even if the power fails. Learn how they operate. Instructions are almost always written alongside the mechanism. Find every escape route available and find any points in the tram that might give you some cover or concealment if you need it.

The operator of the train can often be contacted via a wall-mounted intercom. Intercoms may be a good method of communicating under certain circumstances. Be aware, however, that if you are trying to keep quiet while alerting the train operator to possible terrorist activity, you may expose your intentions when using an intercom. Even if you whisper into the intercom, the train operator may answer loud and clear over the speaker of your intercom. Intercoms are normally loud by design. Keep this in mind when deciding whether to use it or not.

When approaching a station, you should be looking into the station and not down at your newspaper. The best time to spot danger is before it gets aboard your tram. Take a few seconds at every stop to look into the station and at the people and activity in the station. If something doesn't look right, get off right there and don't waste any time. Call the authorities at the nearest phone and let them know what made you uneasy. It could be a tip that saves the lives of everyone on the train that just left the station.

If you do have to exit the tram in an area outside of a designated station, beware of any electrical systems on the tracks or overhead that could come into contact with you. Several hundred volts at a very high amperage flow along the corridors of electric railways. Sometimes this power is carried along cables suspended above the tram. Other times the power is carried on an electric "third rail" that runs alongside the train and is kept in contact with the train by a series of protruding paddles on each car. Do not come into contact with any of these electric surfaces. The power carried along these lines is sufficient to kill you instantly.

Once in the corridor make use of any exits you find. Look for doors along the sides and ladders leading to hatches overhead. If you are being chased or if there are bullets flying, you will need to get around a corner, into a hollow, or just continue putting distance between you and danger by making a run down the tracks. Pay attention to signs along the wall if you can. Many mass transit corridors have arrows showing direction and even the distance to the next exit. Use this information to save yourself.

AIRBORNE TRANSPORTATION

Terrorism aboard an aircraft can present a unique set of complications. One of the most obvious differences is that you are in a vehicle that you may not be able to control even if you are successful in ridding yourself of the terrorists. In the case of high-altitude jetliners, you are in a pressurized compartment that is susceptible to major punctures in the skin of the aircraft, or even an open door. Rapid depressurization at high altitude means that you could have lost most of your oxygen as well. If the oxygen masks fail to operate properly, a mass loss of consciousness could occur affecting everyone aboard, including the pilot. Furthermore, the escaping air from the inside of the aircraft will create a rush of debris and, possibly, people as the wind carries everything in its path out of the hole into the frigid, oxygen-depleted air. And then there's the problem of the 35,000-foot fall.

There are many myths concerning gunshot holes and depressurization. I want to address a common one right now. It is widely believed that a gunshot through the skin of an aircraft cruising at high altitude will cause the aircraft to rapidly depressurize and even disintegrate. The fact is that air escapes from all jetliners all of the time anyway. There are valves that allow air to escape while more air is pumped into the aircraft by other mechanisms. This allows the inside of the aircraft to maintain a relatively comfortable and constant pressure. Doors leak and seams leak, too, because there is no perfect seal in every part of the aircraft. Knowing this, you can imagine that a hole in the skin of the aircraft the size of even a .45 caliber bullet would cause only a momentary loss of a small amount of pressure before the valves and pumps stabilize the pressure again. More holes and larger holes will make the valves and pumps work harder, but any reasonable number of holes will be accommodated by the systems aboard. No, the aircraft won't disintegrate under these conditions. There is actually far more danger from a bullet passing through flight control surfaces, hydraulics, or electronics.

Bombs aboard aircraft can cause catastrophic failure. Not only is the integrity of the skin compromised, the airframe itself is often broken and this can lead to the disintegration and subsequent crash of an aircraft. The destruction of flight control systems could make controlled flight impossible. Even a small explosion aboard an aircraft could be devastating. Admittedly, there is nothing you can do if you have been given no warning and a bomb explodes in the belly of your aircraft. On the other hand, if you get some warning, as was the case with Richard Reid, you may be able to save yourself and everyone aboard the aircraft by acting swiftly.

In this day and age, anyone who acts in a manner that threatens the safety of the aircraft should be apprehended and held by the passengers or the crew at once. There is a very low tolerance for threats aboard aircraft in the post 9-11 world. You will have to be your own judge of appropriate measures to stop a threat. If an attack is imminent or actually being carried out, it is absolutely imperative that the terrorists do not detonate any devices or gain access to the cockpit. Do whatever you must to put an end to the plans of the terrorists, and to them as well, as the opportunity and need arises. When you do engage a terrorist, as always, go hard, go fast, and don't stop. The longer this engagement goes on, the more likely it will be that a disaster will occur. Get it over with as quickly as possible by using overwhelming force and determination.

If your aircraft has crash-landed you should get out of the aircraft immediately. Fires, explosions, and sinking in deep or cold water become immediate threats to your survival. Your chances of survival will greatly increase if you know where all of the exits are on the aircraft. When you first board the aircraft look in front of you and behind you to find the nearest exit. Don't just look at them and make a note. You should actually count the number of seats to the front and to the rear that indicate where exactly the exits are. In a smoke-filled aircraft that has lost electrical power you may not be able to see anything at all. Knowing the number of seats between you and the nearest exit could give you the edge you need to survive a real catastrophe. Be prepared to get down on the floor and crawl the number of seats you know to be between you and the exit, find the way to the door, and get out fast. If the emergency slides have been deployed, you should jump into the center of the emergency slide and move away from the slide the minute you hit the ground at the bottom. Don't waste time sitting down to slide. Sitting down will slow your departure from the aircraft and slow your slide to the bottom. Jump! It's quicker and actually safer than sitting down to ride the slide.

When planning airline travel, it is important to consider the type of clothing you will wear on the aircraft. Consider your clothing from the view of safety over style, though both can be accommodated with a little planning. Don't wear high-heeled shoes, sandals, or open-toed shoes; these types of footwear are not suitable for emergency situations. High-heeled shoes could even damage or deflate the emergency slide that is deployed for rapid evacuation. Leather or canvas shoes that are laced or strapped are best. Wear loose, comfortable clothing that allows for freedom of movement. Avoid wearing synthetic materials like polyester and nylon. In a fire these clothes will surely melt to your skin and could even catch fire. Wear wool, denim, leather, or cotton fabrics. Long sleeves and long pants will offer much more protection than short-sleeved shirts and shorts.

As in every situation, remain alert on the aircraft. The high-altitude, fuel-filled, high-speed, pressurized vehicle in which you are traveling makes early detection of terrorist activity all the more important.

WATERBORNE TRANSPORTATION

Mass transit vehicles on the water are common in bay and river cities. Ferries traveling across bays and rivers are a convenient and rapid mode of transportation and as such they carry hundreds of passengers per trip and thousands of people per day. Seagoing vessels such as cruise liners also carry hundreds, if not thousands, of passengers at a time. This fact may make waterborne modes of transportation targets of terrorism. The Achille Lauro incident in 1985 highlights the fact that terrorists have taken note of the headlines that can be garnered by seizing waterborne craft containing hundreds of passengers.

There are some similarities between waterborne and airborne craft. These similarities stem from the fact that it is a relatively captive group of passengers who are not exactly free to flee from the terrorists. Another similarity is the fact that both waterborne and airborne craft require some amount of skill to pilot. In the case of waterborne vessels, though, the parameters for failure are somewhat more forgiving in most circumstances. One major difference, however, is that you may be able to jump from a boat or ferry, whereas you will not be contemplating jumping from a jetliner flying at 35,000 feet.

The greatest complications hindering your survival, if you decide to jump from a boat into the water, are the distance you will fall before striking the water, and the temperature of the water itself. I have made a jump or two from the flight deck of an aircraft carrier into the ocean 60 feet below. I can attest that if you fail to strike the water properly from this height the experience will be painful. Those jumps were into relatively warm water. The real danger comes when you are forced to jump into cold water. I will address these two issues separately.

First, I would not recommend jumping from a vessel into the water unless there is a clear indication that the water below is more survivable than the conditions aboard the vessel. Stay aboard the vessel if at all possible. If you find it necessary to jump from a ferry or cruise ship, it would be best make your jump when the terrorists are unaware of your intentions and when they are not looking. To drop from any part of the vessel into water that is more than a few feet below you, you need to do it properly. Make as certain as possible that the water is deep enough to avoid hitting the bottom. Always jump feet first with your feet either crossed or squeezed tightly together. Your legs should be straight. Hold your arms down vertically with your hands covering your groin area and cupped tightly in your crotch. (This goes for men and women equally.) Clinch your buttocks tightly and enter the water as vertically as possible. Upon entering the water you should open up your arms and legs, but only *after* entering the water. This will slow your descent through the water and help you to avoid hitting the bottom in shallow water.

If there is any reason to believe that the terrorists will make an attempt to shoot at you after you enter the water, you will have to make a decision about slowing your descent. Obviously, bullets being fired at you will be most effective near the surface, so under these conditions you must go as deep as you can and stay there for as long as you can. Swim while you are underwater so that you don't come up exactly where you were last seen. If the vessel is in motion, try to swim directly away from the sides of the vessel so as to avoid the propellers. Chances are that you will attract some attention when you come up for air, so exhale on the way up and grab another breath of air as soon as you break the surface and then get underwater before the bullets begin to fly. Repeat this procedure for as long as the threat exists. It will be exhausting, but under the circumstances it may be your best chance to survive.

Making a jump into cold water is an even more serious consideration. Hypothermia is a deadly concern in cold water, and the water doesn't have to be freezing to cause hypothermia. In water that is 50 degrees Fahrenheit (10 degrees Celsius) the average person will survive only about 2 hours while treading water with no flotation. Swimming or treading water will actually cause your body to cool up to one-third faster than if you are using flotation and remaining still. When you are in cold water, swim only if you are absolutely certain you can make it to land or some item that you can climb onto to get out of the water. Remember that distances are hard to judge when you are looking across water, and everything will appear to be much closer than it actually is.

If you are able to secure a life jacket or other flotation device, there are a couple of things you can do to reduce the amount of heat you lose. First, you must try to keep your head out of the water. Nearly half of the heat that is lost in your body is through your head and neck region. Next, you can assume a fetal position by pulling your legs toward your chest and wrapping your arms around your knees or crossing them across your chest. This will protect your midsection to some degree and reduce heat loss. If there are several of you in the water you can huddle together by wrapping your arms around each other while forming a tight cluster of bodies. This will allow each of you to share the warmth and increase the insulation provided by the group.

Taking the steps described above will increase your chances of surviving in the water. Again, I must stress the importance of staying on board the vessel if at all possible. Even in warm water, the arduous task of treading water and waiting for help can be exhausting. In a river or bay where the currents are strong, even an experienced swimmer will have difficulty. Ultimately it is your decision whether to remain on the vessel or to take your chances in the water. Consider your options carefully.

If you do decide to remain on a waterborne vessel that has been seized by terrorists, you may choose to remain passive or to fight. I have covered both of these possibilities in earlier chapters. On a waterborne vessel, should you choose to fight, you may have many improvised weapons at your fingertips. Ships and boats are notorious for having lots of sturdy objects that could be used as weapons against terrorists. If you choose to conceal yourself, you may find many hiding places aboard a seagoing vessel or even a moderately sized ferry. Use everything available to you to secure your safety or to defeat the terrorists, whichever is your goal.

Surviving Distributed Agent Attacks

Distributed agents are those substances that cause harm to you through contact with them. I want to familiarize you with some of these substances, how they are spread, and their effects on the human body. In the next chapter I will offer some guidelines on decontaminating yourself should you be unfortunate enough to come into contact with any of these substances.

Terrorists have used distributed agents in the past. The Aum Shinrikyo cult released low quality, poorly distributed sarin gas in a Tokyo subway in 1993, killing 12 people and injuring hundreds. The Rajneeshees sprinkled salmonella on salad bars in The Dalles, Oregon in 1984, sickening over 750 people. These attacks may only be the tip of the iceberg. Dozens of poisoned dogs and rabbits have been found around the camps of Osama bin Laden in Afghanistan and terrorists interrogated by American authorities have described training that included the use of distributed agents.

Distributed agents have an ability to strike at several levels. If distributed effectively, they can injure or kill massive numbers of victims for very little cash investment from the terrorists. They strike fear into the population in general and can shut down a subway or a city for days or weeks. The cost of treatment for victims, cleanup of the contaminated area, and lost productivity of the population can cost hundreds of millions (or even billions) of dollars. These characteristics make distributed agent attacks an economic terror threat without equal.

The effectiveness of a distributed agent is highly dependent upon the method used to distribute it, the environmental conditions, and the agent itself. It is extremely difficult to get high concentrations of lethal agents to properly disperse over a population. This problem may be overcome by distributing the agent in an enclosed area such as a subway or building, or by placing lethal substances in the food chain. Terrorists have proven to be very creative in finding ways to accomplish their missions. We must remain vigilant and look at our world occasionally through the eyes of a terrorist. Only by doing so will we see our own vulnerabilities.

In the following paragraphs I will introduce you to some of the distributed agents that could be used by terrorists. This is not an exhaustive list of agents, but rather a sampling of the more common ones. There are several Internet sites that describe different kinds of agents and their effects on the human body. Search engine words for these sites would include "bioterror", "biological agent", "chemical agent", "decontamination", and "dirty bomb".

You should be aware that there is no substitute for a biohazard suit and gas mask in an environment filled with a harmful distributed agent. On the other hand, carrying this type of equipment with you is not practical on a daily basis. The best protective suit in the world won't help you if you leave it at home when you go shopping. Knowing this, I will direct my efforts to teaching you what to do if you find yourself in an environment where a distributed agent has been released and you do not have immediate access to a biohazard suit.

As an aside, I want to issue a warning to you concerning the wearing of a gas mask. In the event that you have donned a gas mask in an environment filled with a harmful agent your mask may become the target of those around you. The realization that a contaminant is present in the air can drive people to do things they normally would not do. If you are wearing a gas mask and others around you are choking on an agent you would do well to move away from them if you can. People may begin to clamor for your mask and, in their horrified state, may remove your mask in an attempt to save their own lives. The situation is similar to a drowning person being approached by a lifeguard. The person that is drowning often will begin to climb onto the lifeguard in an attempt to get a breath of air, thus putting both of them at risk of drowning.

It is a natural reaction to want to cover your mouth and nose with your hand or a cloth when you see or smell an airborne contaminant. This is a good reaction. Place any mask that you have available over your face if you suspect that a contaminant exists in the environment. If you don't have a mask, use your shirt, handkerchief, or similar piece of cloth to cover your mouth and nose. Be sure that you seal all the way around your mouth and nose and don't allow any air to pass around the cloth. Make the air go through the cloth. The moisture from your breath will make the cloth slightly damp. This will help the cloth become more effective at catching small particles. If you are using a cloth, you can double or triple the cloth thickness by folding it. Just make sure you can still draw air through the cloth. Don't make it too thick or you will be unable to breathe through it. The idea here is to limit the amount of foreign particles that reach your lungs. I should warn you that chemical agents disbursed as a gas will pass through the cloth, and the cloth probably will not be very effective. Having said that, it is sometimes a matter of keeping out just enough of the contaminant, even if you can't keep all of it out. Do your best and don't give up. Anything is better than nothing, and it might just be enough to save your life.

RADIOLOGICAL AGENTS

Radiological agents are substances that cause illness by exposing the victim to radiation. Terrorists who use radiological agents alone are not likely to cause death on a massive scale, but will use the weapon to strike a psychological and economic blow. Even so, the victims of radiological agents are likely to have health problems for years and the contaminated area could be uninhabitable for a long time.

Exposure to high levels of radiation has many health risks. Radioactivity is readily absorbed into the soft and hard tissues of the body. The cells that make up the lining of the intestine, white blood cells, and the cells that manufacture white and red blood cells seem to be particularly vulnerable to high levels of radiation. The early signs of severe radiation sickness are usually nausea, vomiting, diarrhea, and extreme fatigue. In the longer term, illnesses caused by excessive radiation exposure include increased risk of cancer and genetic disorders.

The most obvious sources of radiological agents are nuclear explosions and nuclear power plants. The detonation of a nuclear device or the breeching of a nuclear power plant's core would release enormous amounts of radioactive material into the environment. Radioactive agents in the environment are carried in the wind and through waterways. The radioactive particles carried in the air are collectively called "fallout". There are other methods that a terrorist might use to create radioactive fallout. One of these methods is the use of a conventional explosive (such as dynamite) to vaporize and disburse radioactive materials. Such a device is called a "dirty bomb". A dirty bomb is a radiological weapon, and not a nuclear weapon, because the explosive product does not involve nuclear fission or fusion. Radiological agents for dirty bombs could include strontium-90, americium, beryllium, and cesium-137, among others. Dirty bombs are relatively easy to build, are inexpensive, and the materials are available from a variety of sources.

Possibly the most likely radiological agent to be used by a terrorist in making a dirty bomb is cesium-137. Cesium-137 is widely used in medical equipment, as a treatment for cancer, and in atomic clocks. Keeping track of cesium-137 has been difficult and it has been found inside discarded medical equipment in scrap yards.

Given its ubiquitous nature, it would be easy for a terrorist organization to obtain substantial amounts of cesium-137.

Typically, the radiological agent will not kill you in the short-term unless you are exposed to an extremely high concentration of the agent. Even then, death may not come for days or weeks. Longer-term threats to your life are quite real and your risk of cancer grows with the amount of radiation you absorb. The intensity of the radiation and the amount of time you are exposed to it will determine the extent of damage to your body. The higher the radiation and the longer you are exposed, the more damage your body will suffer. Radiological agents are colorless, odorless, and tasteless. Don't be fooled into thinking that your environment is safe just because you don't see, feel, smell, or taste anything. That mistake can lead to serious injury or death.

Avoiding the fallout of a radiological device is crucial to avoid injury. If you are close to the device when it explodes, and you are able to leave the area, then leave. The direction of your flight should most immediately be directly away from the explosion. Fallout will be more potent near the explosion and will diminish with distance. Within a few minutes or hours the fallout will begin to settle. What you do in this time will greatly affect your chances of surviving this type of terrorist attack.

Information will be critical and having access to a radio would be very helpful. Announcements should be made over the radio telling you where the most fallout will occur and what path you may take to avoid it. Use this information to direct your flight. If no information is available, or if you don't have a radio with you, remember that the fallout will travel with the wind. Look around you for any indication of wind direction. Smoke, clouds, trees and your own skin are indicators of wind direction. Move in the opposite direction of the wind, so long as your movement does not bring you closer to the site of the radiation source. If there is no wind, flee directly away from the source of radiation. If you find that traveling against the wind will only bring you closer to the source of the radiation, then it would be better to travel in a direction at right angles to the wind while moving away from the radiation source. The ultimate goal is to get as far as you can from the radioactive agent source while reducing your exposure to fallout.

Let's say you are standing on the corner and an explosion rocks the top of a building to your southeast. Someone tells you that the authorities suspect that the explosion was a dirty bomb. Immediately you note that the wind direction is from south-to-north. You already know several things at this point:

- There is potential for radioactive fallout.

- The explosion was southeast of you in close proximity.

- The wind is blowing from south-to-north.

- Fallout travels with the wind and settles within a few minutes or hours.

- The fallout is coming toward you because the wind is bringing it your way.

- Traveling directly against the wind will bring you closer to the location of the explosion, which is the source of radiation.

What will you do? You will either seek shelter immediately or you will flee your location traveling northwest and then seek shelter. If this answer was not apparent to you, then draw out the above scenario on a piece of paper to help you visualize the scene and your direction of flight.

Shelter from a radiological agent is absolutely essential. You must be able to identify effective shelter and get into it as quickly as possible. The amount of protection you receive from the radiation will depend upon the type and thickness of the material surrounding your shelter. Wood, sheetrock, glass and fiberboard are poor protection. Concrete, metal, and soil are good protection and the thicker it is, the better. Protection from radiation will be stronger in the center of a building than on its edges. Reinforced concrete and steel buildings, basements, and sealed underground parking garages make good shelter. Try to find a shelter that has no windows, or at least as few as possible. If you are outside, crawling into a ditch or tunnel will provide some protection as well. Bridge and overpass abutments may not be optimal, but they are better than nothing. Be alert for anything that could give you some protection and get into it quickly.

When moving from one place to another in an area of fallout try to cover yourself as much as possible to reduce the level of contamination on your body. Raincoats are good and ponchos are better. Wrap yourself in anything that can protect you: garbage bags, plastic sheets, and blankets can be useful. Breathe through a wet cloth like your shirt or handkerchief to keep the particles from entering your lungs. Don't eat or drink anything that may have been exposed to radioactive materials. Minimize the time you spend in the open; time is critical.

If you are in a car and you can maneuver, then do so quickly. Shut your windows and vents, turn off any air conditioning or heating, and keep your doors closed. Don't get out of your car until you reach substantial shelter. Be mindful of your fuel situation, too. Don't pass up good shelter only to run out of gas in the open without shelter.

Remove your clothes before entering your shelter, and don't take anything from the outside into the shelter with you. Around ninety percent of the fallout that collects on you will be on your clothes. Get rid of them. If you can't remove all of them, then at least remove the outer layers. You should decontaminate your body before entering your shelter if at all possible. If you must bring clothes or items into your shelter then you should decontaminate them as well. I will cover decontamination procedures for your body and your clothes in the next chapter. Turn off all ventilation and seal any vents, windows, and doors. Use plastic sheeting and duct tape if you have it. If plastic sheets and duct tape are unavailable use anything at hand to block the flow of air. Once you take shelter you should remain there until you are notified to leave by authorities or you are forced out by environmental conditions. You may be there for several days, so don't panic, and accept the fact that you are doing exactly what you should be doing. If you have safe food or water in your shelter you should be very conservative with consuming it. You could be in your shelter for a long time. Impatience in a radioactive environment can kill you. Be patient.

CHEMICAL AGENTS

Chemical agents are concentrations of chemical substances specifically designed to impair, injure, or kill victims in a variety of ways. Chemical agents range from simply manufactured concoctions to exotic agents that require sophisticated laboratories to produce.

Chemical agents are disbursed as solids, liquids, aerosols, and gasses. An aerosol is a substance that is distributed as tiny droplets in air. Hair spray and spray air fresheners are examples of aerosols. Gasses are sometimes called vapors, and are actually individual molecules of a substance suspended in air. Steam and automobile exhausts are examples of gasses.

Chemical agents have different levels of persistency, meaning that some may "hang around" longer than others after they have been released. A substance that evaporates quickly, like water or alcohol, has a low persistency because it doesn't stay in the environment for very long. A potentially deadly side effect of substances with low persistence is the fact that the evaporation process that some undergo actually can create lethal gasses. Substances that cling to surfaces and do not evaporate quickly, like oils and syrups, have a high persistency and may be around for days. A terrorist may mix a thicker, inert substance with a chemical agent to cause the chemical agent to be more persistent, making it harder to clean up and more likely to come into contact with more victims.

Victims of a chemical agent may show symptoms immediately, or symptoms may take days to appear. The rapidity with which a chemical agent victim shows symptoms is dependent upon the person to some extent. Each chemical agent will attack in a slightly different way and terrorists may choose a particular chemical agent for its specific properties. In some circumstances, terrorists may choose to use chemical agents that attack their victims quickly, injuring or killing them within minutes. In other circumstances, terrorists may choose to use chemical agents that delay their symptoms, thus allowing possibly thousands of people to become victims over a matter of days before anyone is aware that a chemical agent attack is happening. Imagine a delayed chemical attack in a busy subway where the chemical agent remains undetected for two or three days. Suddenly the hospitals are packed past capacity with thousands or tens of thousands of victims that begin to show symptoms. This type of attack would effectively shut down a city for some period of time and the medical treatment, loss of workers, and cleanup would cost hundreds of millions of dollars.

Chemical agents can be grouped into two types of desired effects: Non-Lethal and Lethal. Non-Lethal or "harassing" chemical agents like tear gas and pepper spray are designed to cause discomfort or incapacitate you so that you are unable to fight or resist as effectively as you normally would. Harassing agents cause discomfort within seconds and generally target the skin, lungs, and mucous membranes. Symptoms may include choking, burning eyes and skin, vomiting, and mucous discharge through your nose. If you have been exposed to pepper spray or tear gas, don't panic. The effects are temporary though uncomfortable. If you are in a position to get out of the contaminated environment, then do so. Hold your arms away from your body and allow the chemical agent to evaporate. Breathe normally to clear your lungs and allow tears to clear your eyes. Do not rub your skin or eyes. If you are unable to remove yourself from the environment, or you are engaged in resisting the terrorist, just remember that pepper spray and tear gas are not lethal. Continue to do what you must to the best of your ability.

Lethal chemical agents are designed to cause severe injury or death. Lethal chemical agents may kill within seconds or minutes, while others may take hours or days to kill their victims. Lethal agents typically are divided into four groups: choking agents, blister agents, blood agents, and nerve agents.

Choking Agents

Choking agents usually attack the lungs and windpipe by causing a profuse flow of fluid from the membranes that, in turn, drown the victim in his own fluids. Most of the time the victim has inhaled the choking agent as a gas or aerosol. Some examples of choking agents are chlorine, phosgene, diphosgene, and chloropicrin. None of the agents named are very persistent and, as a result, are not in the environment for very long after being released. Victims of chlorine and chloropicrin gas usually show symptoms within minutes, while the symptoms of phosgene and diphosgene may be delayed for up to twenty-four hours.

If you suspect you have been exposed to a choking agent you must immediately seek medical attention. Unfortunately, at the time of this writing there exists no antidote for choking agent poisoning. Medical personnel may be able to provide support through mechanical breathing equipment or by clearing the lungs of fluid. If you are unable to reach medical professionals you should immediately decontaminate yourself and seek shelter.

Blister Agents

Blister agents, also called vesicants, damage tissue in painful and sometimes horrifyingly grotesque ways. Any tissue that comes into contact with a blister agent is invaded within the first couple of minutes, so decontamination of the agent may have little effect after this period of time. Most of the blister agents are the consistency of light oil and generally are quite persistent in the environment. Typically blister agents are disbursed as a liquid or an aerosol, and attack skin if touched and the respiratory tract if inhaled. The blister agent irritates the affected area and kills the cells that make up the tissue. The destruction of the tissue and the infection that follows is what gives the "blistering" symptom. Blistering agents are listed as lethal chemical agents because an inhaled blistering agent can damage the lungs so severely that untreated victims may die. It is more likely, however, that victims will not die from the exposure but rather will suffer extremely painful blisters and possibly blindness. Some examples of blistering agents are sulfur mustard, nitrogen mustard, lewisite, and phosgene oxime. Sulfur mustard and nitrogen mustard agents are quite persistent in the environment and their symptoms are delayed from one to four days. Lewisite is relatively persistent and the symptoms appear within minutes or hours. Phosgene oxime is not very persistent but extreme pain and tissue damage appear immediately.

If you suspect that you have been exposed to a blister agent you must seek medical attention immediately. Blister agents do their damage quite quickly, even though the symptoms may not appear for hours. If medical help is not immediate, then you must decontaminate your skin and remove your clothes as quickly as possible. In the vast majority of cases blister agents are not fatal though they are very painful and may cause long-term health problems.

Blood Agents

Blood agents are, as the name implies, distributed to tissues and organs through the movement of blood. Blood agents are usually disbursed as a gas and the agent enters the blood through the lungs. The lethal quality of a blood agent is in its ability to inhibit cells from transporting and utilizing oxygen, which has the effect of suffocating the victim. Examples of blood agents include hydrogen cyanide, cyanogen chloride, and arsine. None of the blood agents mentioned here are persistent in the environment. Hydrogen cyanide and cyanogen chloride act quickly, usually within seconds to a few minutes. Arsine is sometimes described as having a garlic-like smell and its symptoms are commonly delayed for several days.

If you suspect that you have been exposed to a blood agent, you must immediately seek medical attention. Military and some first responder units may carry ampoules of amyl nitrate or other blood agent antidotes. If you are unable to reach medical professionals you must immediately decontaminate yourself and seek shelter. Blood agents can be quite deadly in concentrations, but victims of blood agent attacks who survive until the agent dissipates have a chance at full recovery.

Nerve Agents

Nerve agents disrupt nerve cell communication. Most often, nerve agents are absorbed into the lungs by inhalation or directly through the skin. Nerve agents are capable of producing violent and painful death with extremely small doses. Examples of nerve agents are tabun, sarin, soman, cyclosarin, and VX. All these nerve agents act on the body very quickly. VX is the deadliest nerve agent in existence and only a few milligrams touching the skin can cause rapid death. VX, with the consistency of motor oil, is very persistent and may contaminate an area for weeks. Tabun and sarin are not very persistent, while soman and cyclosarin hang around in the environment for a while longer.

If you have no warning and become the victim of a nerve agent attack, you must seek medical attention immediately. Military and some first responder units carry injections of atropine and pralidoxime chloride that can mitigate some of the nerve agent symptoms. Valium has been used to reduce the convulsions that come with the onset of nerve agent poisoning. Only qualified persons should administer antidotes and only after a positive confirmation has been made that the victim is indeed suffering from nerve agent poisoning. If you are unable to reach medical professionals immediately, you must decontaminate yourself as quickly as possible and move to shelter.

If you have some warning that a chemical attack is imminent, then seek shelter immediately. When moving to shelter, cover as much of your skin as possible with clothing, raingear, ponchos, plastic sheeting, or bags and move quickly. Breathe through a moistened cloth if you don't have a gas mask. Remove your clothing before entering your shelter, or decontaminate your clothing before allowing it inside the shelter. Decontamination of skin and clothing is discussed in the next chapter. Chemical agents pose a special risk if they come into your shelter because the gasses produced by simple evaporation (from your clothes) could kill you. Don't take any chances; leave the clothes outside your shelter. Sheltering is discussed in detail in the chapters titled *Managed and Ready Shelters* and *Finding and Creating Improvised Shelters*.

BIOLOGICAL AGENTS

Biological agents are either living organisms or derivatives of living organisms that can cause serious injury or death. Biological agents are generally divided into three categories: viral agents, bacterial agents, and toxins. Viral and bacterial agents are living organisms that can reproduce and have the unique ability to actually increase environmental contamination over time. Biological agents are sometimes classified as either stable or unstable. A stable biological agent would be able to live in the environment for long periods of time, possibly years, after being released. An unstable biological agent doesn't survive for very long in the environment after being released.

Viral Agents

Viral agents do their damage by infecting a person with a few viral organisms, usually through inhalation or ingestion. Once inside the body the virus invades a few cells and begins replicating itself inside each cell. After the cell is inundated with thousands of copies of the virus, the cell bursts and releases the new copies of the virus into the bloodstream and surrounding tissue. Each of these new copies of the virus goes on to repeat the procedure, this time in thousands of cells, which eventually burst with new copies. Soon the victim's body is overwhelmed with the virus and serious illness or death follows. Examples of viral agents that may be used by terrorists include smallpox, Venezuelan Equine Encephalitis, Ebola, and Marburg. Smallpox is very stable in the environment, can spread as an aerosol, and is easily transmitted from person-to-person if the population is not immunized against smallpox. These qualities, coupled by the fact that it can severely disfigure its victims while killing about thirty percent of them, makes it an attractive biological agent for terrorists. Venezuelan Equine Encephalitis (VEE) can be spread as an aerosol but it is not considered very stable. The characteristic that makes VEE a potential biological agent for use in terrorism is that less than one hundred organisms are needed to infect a victim. VEE causes severe illness and kills around ten percent of its victims. Ebola and Marburg viruses fall into a class of viral hemorrhagic fevers that cause high fevers and leakage from blood vessels. Victims eventually begin bleeding from eyes, ears, nose, mouth, and ultimately through the skin as well. Ebola and Marburg are not stable in the environment and may be disbursed as an aerosol. The virus can spread from person-to-person through direct contact with the body fluids of an infected person. Ebola and Marburg viruses are deadly, killing ninety percent of their victims.

During the writing of this book, a new viral threat emerged in China and has since swept across the globe. Severe Acute Respiratory Syndrome (SARS) has infected thousands and killed hundreds. The highly infectious nature of SARS makes it especially difficult to control without massive quarantines. The SARS virus also seems to be persistent on surfaces for at least several hours, making decontamination a priority in public places. Travel restrictions, loss of workers, medical treatment, and decontamination efforts are costing billions of dollars. I wouldn't be surprised if SARS becomes a future bioterror weapon. Imagine if a group of "suicide spreaders" intentionally infected themselves with the SARS virus and then came into the United States before symptoms began to show. These suicide spreaders could enter the population and mingle with the crowds in mass transit terminals while coughing, sneezing, and wiping the agent expelled from their bodies onto people and surfaces alike. The economic and medical toll could be devastating.

Bacterial Agents

Bacterial agents, like viral agents, do their work by infecting their victims with a few bacterial organisms. The bacteria may enter the body through inhalation, ingestion, or breaks in the skin. Once inside the body the bacteria begin to multiply by the millions. They do not invade cells as viruses do, but rather cause damage to the body by excreting toxic substances that overwhelm the body, eventually causing severe illness and death. Most bacterial agents are very stable in the environment. Examples of bacterial agents that might be used as weapons of terror are anthrax, pneumonic plague, and tularemia. Anthrax may stay dormant in an environment for years as spores. Anthrax infections of the intestine or lungs are very dangerous and left untreated will kill a high percentage of those infected. Skin anthrax is serious but not as lethal. Anthrax is not easily transmitted by person-to-person contact. Pneumonic plague is an extremely serious infection of the lungs and may be spread from person-to-person by breathing the aerosol droplets from the breath of an infected person. The fatality rate for untreated victims of pneumonic plague is over ninety percent. Tularemia is less lethal than anthrax or pneumonic plague but is highly infectious, requiring as few as ten organisms to cause infection. The organism that causes tularemia is very stable and may contaminate an environment for weeks in moderately moist areas. Person-to-person transmission of tularemia is unlikely.

As you can see from the descriptions above, detecting biological agents before you have been exposed is difficult. It is more likely that you will be informed of a biological attack by civil authorities after the medical community has noticed a pattern of illness. If you suspect that you have been exposed to a biological agent you must immediately seek medical attention. In the case of bacterial agent exposure medical professionals have had some success in treating victims with antibiotics. The sooner the treatment starts, the more likely the victim is to recover. If you are unable to reach professional help, you should immediately decontaminate yourself and seek shelter from the biological agent.

Toxins

Toxins are poisonous substances produced by or derived from living organisms. Most toxins are stable in the environment and typically can be disbursed as aerosols, ingested by food or drink, or injected directly into tissue or blood. Examples of toxins that could be used by terrorists include ricin and botulinum toxin. Ricin is an extremely toxic substance that is derived from castor beans, which are grown worldwide. As little as one milligram of ricin could kill an adult, and death usually occurs within three to five days. Ricin can be made cheaply and easily, which could make it attractive to terrorists. There is no known antidote for ricin poisoning. Botulinum toxin is derived from a particular genus of bacteria and it is one of the most poisonous substances known. Persons poisoned by botulinum toxin may begin to show symptoms in as little as twelve hours, but it is possible that symptoms may not occur for up to several days. Left untreated, botulinum toxin will kill a large percentage of its victims.

If you find yourself in an environment that you suspect contains an aerosolized toxin you should breathe through a cloth and cover yourself as completely as you can while making your way to shelter. Toxins do not penetrate skin very effectively, but be mindful of any cuts you may have and protect them from exposure to the toxin. Use clothing, bandages, or even apply tape directly to the cut to seal it from the environment.

A terrorist may use a toxin somewhere in the food chain. The Rajneeshees in The Dalles, Oregon sprinkled salmonella toxin on salad bars. The reason that this was so effective is that people consume items from the salad bar without cooking them. Heat sufficient for cooking foods will inactivate many toxins. It is probably a good idea to avoid salad bars altogether. Vegetables bought at the store should be thoroughly washed and scrubbed before eating. If you are eating out, it is safer to eat only cooked foods.

Decontamination is not applicable for toxins that are directly injected or ingested since the poison is not being absorbed or inhaled. For aerosolized toxins, however, decontamination should be performed immediately if medical attention is not available.

Decontamination

For the purpose of our discussion here, decontamination is defined as the act of removing or neutralizing a harmful agent from skin, eyes, or clothing. There are other procedures that may be used to decontaminate your body internally, and steps for decontaminating equipment, but I will not be discussing these methods in this text. I want to concentrate on removing the harmful agent from you and your clothing and getting you to some sort of shelter as quickly as possible.

SKIN DECONTAMINATION

Some general principles apply to all decontamination procedures. These principles are:

- When dealing with a life-threatening injury, take care of the injury first and consider decontamination a lower priority.

- Remove yourself from the contaminated environment before beginning decontamination procedures.

- Remove all clothing before beginning decontamination of skin and eyes. Remove clothing by cutting away from your body if possible to keep from contaminating your face while pulling clothing over your head.

- Decontaminate yourself outside your shelter, and do not allow contaminated persons, clothing, or objects into your shelter.

- When decontaminating, begin with the highest point of your body and work down.

- Stand on a raised platform, such as a chair if it is available, and always allow the decontaminant to stream down and away from your body.

- Scrub skin very lightly and do not abrade the surface or cause the skin to become reddened and irritated.

- Avoid washing agents into skin breaks and openings such as cuts, abrasions, eyes, nose, ears, the mouth, genital, and rectal openings.

- Pay special attention to hands, fingers, and fingernails.

- Be thorough and work one section of the body at a time beginning with the highest point.

- In the case of long hair or styles (such as dreadlocks) that may be difficult to wash thoroughly, cut the hair close to the scalp and discard it.

- Do not allow agents being washed from hair to run into your face or over your shoulders. Tilt your head back and let the agents fall away from your body while washing.

- When helping another person decontaminate, or if you are decontaminating an item, use gloves, eye protection, a face shield, and any other appropriate and available protection.

Radiological agent decontamination is straightforward. Radiological agents settle on clothing and skin when you are in a contaminated environment. The penetration of clothing and unbroken skin is usually minimal with these agents. By removing your clothing you will remove a significant amount of radiological agents from your body. Once removed, the clothing should be packed into a plastic bag and stored outside your shelter.

Decontaminate your body by using generous amounts of mild detergent in lukewarm water to wash your hair and skin. Shampoo, liquid soap, bar soap, bubble bath and other detergents are adequate for making a decontaminant solution so long as the concentration is kept low. Vinegar can also be used to remove radiological agents. You may scrub with a sponge, soft brush, or cloth. If using a brush, be sure the brush is soft, and scrub very lightly to avoid irritating the skin. Rinse thoroughly with lukewarm water. If someone is available to help you wash those hard-to-reach places then let them help. This is no time for modesty. Pat your skin dry with a clean towel if you have one; otherwise, allow yourself to dry in the air. Don't reuse any of the decontaminant solution that was flushed over your body.

A large amount of decontaminant solution washing over your body is best, but this may not be possible in an emergency. If you find you are low on decontaminant solution or water, improvise by pouring small amounts of the solution onto a sponge (or cloth) and wiping your body with it. Do not dip the sponge into the decontaminant solution. Pour the solution onto the sponge in small amounts and wipe the sponge over your skin in one direction. Work over your body in sections and be meticulous. This may be the most important sponge bath you ever take.

Chemical and biological agent decontamination is a bit trickier in some ways than it is for radiological agents. The reason is that many chemical agents penetrate the skin and clothing very effectively and quickly. Also, some chemical agents are not easily removed with soap and water, and some biological agents are not neutralized with soap and water. When soap and water is the only thing available, then follow the instructions given above for radiological decontamination. This may be sufficient for a majority of chemical or biological contamination cases.

If you have been exposed to a chemical or biological agent, get out of the contaminated environment as quickly as possible. Remove your clothes and seal them in plastic bags if possible. You should be aware that phosgene oxime penetrates plastic and rubber easily. Clothing exposed to phosgene oxime can be dangerous even through a closed plastic bag. Any clothing that has been exposed to chemical agents that evaporate rapidly may produce deadly gasses. Don't allow any clothing inside your shelter that has not been completely decontaminated.

Use flour, talcum powder, or fuller's earth to absorb chemical agents on the skin, especially if the chemical agent is oily or sticky. Oily chemical agents are difficult to wash from skin and the use of absorbent powders will help remove some of the chemical agent before washing. If you don't have any of these absorbent powders, don't worry. Just move on to washing with the soap and water solution described above, or the hypochlorite solution described below. If you suspect that you have been exposed to a biological agent, it might not be worth your time to use absorbent materials. Move on to the washing stage.

Most chemical and biological agents are susceptible to being neutralized and removed by a hypochlorite solution. An optimal concentration of hypochlorite solution for your skin is 0.5%. Most household bleach is between 5% and 6% sodium hypochlorite, which is about ten times the strength you need for your skin. A useable decontamination solution for your skin can be made from one part household chlorine bleach and nine parts water. This means that one gallon of household bleach will combine to make ten gallons of hypochlorite decontamination solution. The solution will smell similar to the water in a chlorinated swimming pool. Use the hypochlorite solution to wash your skin in exactly the same manner described for decontamination using soap and water above. Be especially careful to avoid the eyes with the hypochlorite solution. The hypochlorite solution itself will emit gasses that may irritate your respiratory tract and eyes. Try to avoid breathing the fumes as much as you can, but decontamination is your primary concern.

EYE DECONTAMINATION

Eye decontamination should be accomplished by flushing your eyes with plain water for at least fifteen minutes. If you are wearing contact lenses, remove them. Turn your head to the side and open the lids of the eye closest to the ground. Do not use your fingers or gloves to open your eyelids if there is any chance that your fingers or gloves are contaminated. Pour water slowly from a container into the corner of the eye closest to the bridge of your nose. The water should flush over the eyeball and flow out of the corner of the eye closest to your ear. Do not pour water into your eye in such a way that the water being flushed from one eye flows into the other eye. When you have completed one eye move to the other eye, tilting your head to the opposite side and repeating the same steps as before.

CLOTHING DECONTAMINATION

Clothing decontamination can be accomplished with soap and water or a concentrated hypochlorite solution. Clothing that is contaminated with a chemical agent can continue to give off deadly gasses until it is properly decontaminated. Don't breathe near, or come into direct contact with, contaminated clothing. If you are in a position to bag and seal contaminated clothing then do so. If you must keep your clothes because of environmental or protective considerations you will have to decontaminate them. Use plenty of soapy water if you have it. Better yet, soak your clothing in full-strength household chlorine bleach, right out of the bottle, for one hour. If you are using chlorine bleach at full strength, be sure that you do not use the full-strength bleach for decontaminating skin. Keep the two solutions separate and mark them. Scrub your clothing with a brush if possible. Avoid splashing harmful agents on your body or into your eyes. Be aware that using a full-strength hypochlorite solution is likely to bleach your clothes severely, but in this situation this should not be a consideration. Wear protective suits, if they are available, when handling contaminated clothing.

Rinse your clothes well and let them dry before wearing them or bringing them into your shelter.

Managed and Ready Shelters

Throughout this book I have made reference to sheltering yourself from various threats in your environment. This chapter and the next chapter, *Finding and Creating Improvised Shelters*, will discuss the different types of shelters and the qualities that each brings to aid in your survival. Shelters are simply temporary protective enclosures or places that provide refuge from the environmental threat you are facing. In the following pages I will discuss the different types of shelters, how to make shelter for yourself, and special precautions concerning entering and sealing shelters.

A managed shelter is a permanent, maintained, and stocked shelter designed to protect you in the event of a nuclear, radiological, biological, or chemical terrorist attack. Managed shelters are generally maintained and supervised by government agencies or those commercial industries that manufacture hazardous materials. Even though managed shelters may serve other functions when not being used as a shelter, they have been designated and certified as shelters by the government or industry in charge of them. Managed shelters may also be good blast shelters if they were originally designed with this threat in mind. Check with local officials for the location of managed shelters in your area. A ready shelter, on the other hand, is a room or space that normally serves another function and is not truly a shelter until some effort is made to seal it. A ready shelter is a space chosen by you and prepared by you as an individual. A ready shelter is maintained and stocked just as you would expect in a managed shelter. The ready shelter, however, is not sealed when you first enter it. You will take the last step of sealing the shelter when the threat has presented itself. Ready shelters are convenient and appropriate for home and office spaces where a permanent managed shelter is not possible or practical. Ready shelters are generally designed for shorter-term use than managed shelters. The determining factor is usually in the regulation of human waste disposal and the filtration and exchange of air. For the remainder of this chapter I will concentrate on selecting, preparing, sealing, and supplying ready shelters.

SELECTING AND PREPARING READY SHELTERS

For your ready shelter you should select a room inside your home or office that does not have windows, or at least as few windows as possible. Beware of rooms with false ceilings, especially in the office. False ceilings may allow the passage of air (and contaminants) if the walls do not go all the way to the structural ceiling.

Ideally the room should be large enough to accommodate the number of people you would expect to occupy the shelter for a particular amount of time. Since most of you will not be constructing a new room just to create a shelter, I believe it would be better to offer a calculation that would help you approximate the amount of air available in a room of a given size. Say your basement is 20 feet wide, 27 feet long, and has a ceiling 8 feet high. Just multiply the width by the length by the ceiling height to get how many cubic feet are in the room. In this hypothetical instance, your basement is 20' x 27' x 8' = 4,320 cubic feet. One person uses about 3 cubic feet of air per minute, which is 180 cubic feet per hour. So to calculate how many hours of air are available you divide 4,320 cubic feet total by 180 cubic feet per hour and get 24 hours of air. So your hypothetical basement, if completely sealed, would provide a full one-day supply of air for one person. The same basement will support two people for 12 hours, and three people for 8 hours. As you can see, air is a resource that must be considered in a sealed shelter. If you don't have a single room that would allow the number of people and time that you desire, you might want to prepare more than one room for ready sheltering.

Locate all vents, windows, doors, and any opening that could allow the passage of air. Measure and cut plastic sheets for every opening you find. The sheets should be able to cover the whole opening plus an inch or so of overlap with the wall. Label each one clearly to indicate the opening for which it is intended. For doors, you will use duct tape directly on the seams around the door. Make sure you can seal the door to the threshold at the bottom. If the gap is too wide between the threshold and the door you may need to create something that will fill the gap before taping it. Clearly mark all items so that you and others will know where each item belongs. You can never have enough duct tape, so store several rolls. It can take a lot of tape to secure all of the plastic sheeting, cover any outlets, and seal the door. Store these precut sheets and duct tape in a cabinet or container clearly marked so that it can be readily identified.

Terminal vents (those that blow air into the room) can have enough force to tear the plastic sheeting and duct tape away from the wall and vent, thus allowing contaminated air into your shelter. Return vents (those that take air out of the room) can create a bit of a vacuum in your shelter and cause contaminated air to be pulled into the room through cracks under the door and around windows. It is best that you shut down the ventilation system at the source in addition to sealing vents with plastic sheeting and duct tape. This is also why you should have plenty of duct tape. You may need to repair or reinforce your shelter seals.

Place a to-do list in your shelter in a readily accessible location. This list should contain any tasks that must be done before and during the sealing of the shelter. Instructions for turning off water, ventilation and any other special tasks should be listed. Write on this document the important names and numbers of emergency agencies, relatives, and friends that you may wish to contact after you have sealed your shelter. Be sure to make the to-do list portable. It is easier to follow instructions that you can take with you than it is to try to memorize them. In an emergency your memory may fail or you may skip steps in your task. Make the instructions clear and concise. Type them out or print them clearly so that any person using the shelter will know what to do. You may be elsewhere or injured, so be clear in your explanations.

SEALING READY SHELTERS

Before entering the shelter you should shut off all ventilation to the shelter where applicable. Turn off all switches and thermostats that could cause the ventilation fans to turn on. Close any windows that could provide a further barrier. This prevents airborne contaminants from reaching the outside of your shelter. For example, even if your shelter is in an inner portion of your house without windows, it would still be beneficial to shut the windows of your house to keep the distributed agent from reaching the interior of the house as much as possible. Any barrier is better than no barrier. Shut the windows.

Sealing a ready shelter should be accomplished as quickly and thoroughly as possible. If you are in a shelter where other occupants are expected, but haven't yet arrived then you'll have to weigh the value of sealing the door. If the door is not exceedingly leaky then you may be better served by simply closing the door while waiting for others to arrive. If there is an opening at the bottom of the door, block it temporarily with a cloth. Use this time to go about sealing windows and vents. Once everyone has arrived you may seal the door more securely.

When sealing windows and vents be sure that your plastic sheeting covers the vent or window completely and stretches onto the wall for an inch or two. If you have precut and marked sheets be sure to put them where they belong. Start with large or leaky windows, vents, and openings first. Place the duct tape half on the sheeting and half on the wall. Press the duct tape firmly in place.

Seal the door when everyone has arrived or when you can wait no longer. Use duct tape directly on the seams around the door facing. Tape the area where the door itself meets the facing as well. Plug or tape over any keyhole. Completely block the space below the door with any device you may have constructed for that purpose. Duct tape may not stick well to some types of carpet, so be creative and use a little extra care to seal this opening. Lock the door when you seal it.

Next you should move to seal any electrical, phone, and outlets not in use. Some of these, depending on construction, can let in quite a bit of air through the outlet and around the casing. Cover the entire outlet and facing with duct tape or plastic sheeting and duct tape.

SPECIAL CONSIDERATIONS FOR DISTRIBUTED AGENT ATTACKS

When you are taking shelter before a distributed agent attack has happened, or before a harmful agent has reached the shelter, you have a huge advantage. Assuming that you and your clothing are not contaminated (or at least have been decontaminated), you can enter your shelter without the danger of contaminating it. Your goal should always be to make it to your shelter before the agent does.

The situation is more complicated when you have arrived at the shelter after a harmful agent has reached the outside of the shelter. (This applies only if the shelter doors and windows are closed. Otherwise, the agent is already in your shelter and things will be difficult for sure. Always strive to keep your ready shelter doors and windows closed when you are away.) If the environment outside of your shelter is contaminated, then you and your clothing are contaminated. This means that you must decontaminate yourself before you move into your shelter. An alternative could be to enter the shelter and then immediately remove your clothing and discard it outside of the shelter, but you risk contaminating your shelter to some extent. It is impossible to thoroughly decontaminate yourself while standing in a contaminated environment. This dilemma is why it is so important to reach your shelter before the harmful agent does.

The following discussion assumes you do not have a multistage decontamination airlock system like those used by professional and government agencies. I'm fairly confident that most of us will be trying to get into an inner sealed room in our home or workplace where such elaborate systems are not available. How you enter the shelter, if you do at all, could determine whether your entry will compromise the safe area inside of the shelter.

If you don't reach your shelter before the environment becomes contaminated around it, hope is not lost. You just have to make some decisions and act with extra care when entering the shelter. You are entering the protective shelter from a contaminated environment. Every time you open the shelter door to the contaminated environment you are allowing a certain amount of contaminant into your shelter. This can be a deadly exchange depending upon the type of agent, the concentration of the agent in the environment, and the amount of agent that enters your shelter.

You should first determine if anyone else is in the shelter when you get there. This knowledge may help you determine whether you chance entering the shelter or not. If entering the shelter will endanger the lives of those in the shelter you may want to consider staying outside. Use your time wisely and either try to secure some alternate shelter for yourself, or you could strengthen and seal the shelter from the outside for those inside the shelter. Be aware, however, that staying outside of the shelter unprotected in a contaminated environment to strengthen the shelter for others may prove fatal for you. The decision is yours.

Use common sense when moving into your shelter and try to determine the way that will allow the least amount of contaminated agent to enter. Minimize the time you have the door open. If more than one person is going to enter the shelter within a short period of time it is better to wait and to open the door once for all of them rather than opening it once for each of them. When you enter your shelter under these circumstances you must make an attempt to decontaminate yourself immediately upon entering the shelter as best you can. Throw your clothes out of the shelter and don't allow them to stay in your shelter even for decontamination purposes.

SHELTER SUPPLIES

A ready shelter should be stocked with enough supplies to care for everyone that will be occupying the shelter for as long as two weeks. Two weeks is optimal, but I would consider three days to be minimum. When calculating the amount of water, food, and clean air don't forget any special needs of any of the intended occupants. Infants, children, elderly, and physically or mentally challenged occupants may have certain nutritional, medicinal, or physical support requirements.

An adequate air supply is essential and I have discussed the calculations for air availability above. Since you may not have enough air in one room to accommodate the number of occupants for a significant amount of time you may have chosen to make several separate rooms ready shelters. Keep this in mind as you supply each room. Group the right supplies with the right rooms and make sure that every person knows which room they are to occupy in an emergency. Mix children and people with special needs among the adults and caregivers. Put someone in charge of each room where appropriate. Each room should have a capable, knowledgeable individual that will be responsible for sealing the shelter and managing supplies and hygiene in the room.

When considering the amount of air you must have available for several persons, it becomes apparent that you will probably not have the space available to accommodate occupants for even three days. A two-week air supply is almost impossible without extensive filtration measures. Don't despair over this problem. Most distributed agents dissipate quite rapidly and their ability to harm you will probably be greatly diminished within a few hours. Concentrate on providing several hours or a day of available air. Have a clock, calendar, and radio available in your shelter. These items can guide you when deciding whether to open a window to allow air exchange. Even after the air is clear you will need to provide your own water, food, and shelter for some time. Concentrate on these items for storage in your shelter.

In some instances you may need to communicate with persons in another part of your house that are in separate rooms. It is a good idea to have walkie-talkies to allow you to communicate between rooms if you are out of voice range. Store these walkie-talkies in each room and have extra batteries for them in each room as well.

Pets are an important part of many families, and some pets are practically considered to be family members. Remember to include non-perishable food and extra water for your pets. For air management, count large dogs as one person and small dogs and cats as one-third person each. Plan your shelter resources according to the specific needs of your pet and don't forget to plan for waste disposal. Cats can usually be left in a sealed room without humans. Dogs, on the other hand, have an incredible ability to chew through household materials. If left alone a dog may try to chew or dig his way out to be with you. Worse yet, he may come to your shelter and try to dig or chew his way in, thus breaking the seal to your shelter. Keep him with you if at all possible to avoid both problems.

Ideally you should store one gallon of water for each person occupying the shelter for each day you expect to be dependent upon the stores in the shelter.

Two weeks of water supply is ideal, but you should store at least three days supply of water as a minimum. One gallon per person per day is best, but two quarts per day will do under normal conditions. Remember that children and nursing mothers will require more water. A family of three would ideally have forty-two gallons of water stored in their shelter to last for two weeks.

Food stores should include items that you and members of your family will eat and are familiar to them. Familiar foods can reduce stress in emergency situations. You will know how much food your family eats per day, but a good number for estimation is a minimum of two thousand calories per day of nutritionally balanced food. Five thousand calories would be even better for an adult. This is no time for a diet. Energy and a strong immune system are your main goals. By nutritionally balanced I mean to say it should contain a distribution of around twenty percent protein, thirty percent fat, and fifty percent carbohydrates. I would advocate foods that are high in nutritional value that do not require special preparation or heating. Ready to eat (not condensed) soups are ideal foods for emergency situations. Soups provide both nutrition and moisture in a package that can be stored for up to a year. Soups are usually a good balance of carbohydrates, protein, fat and moisture. If your water supplies are low, you should avoid eating too many foods high in protein and fat that are also low in moisture since your body will require more water to process it. Salty foods, too, will increase your thirst. Canned vegetables and fruits are good for shelter storage. Vegetables and fruits provide high carbohydrates and when canned they also provide lots of moisture. Canned juices are also good sources of carbohydrates and moisture. Canned meats are high in protein but may not have a lot of moisture. Store seasonings that you like including salt, pepper, oregano, Cajun spices, and anything else that can make your food more pleasurable. When stocking foods for your shelter you should carefully avoid any foods that have caused you or any member of your family problems in the past. Such foods would include items that have caused diarrhea. In an emergency situation diarrhea can be at least uncomfortable and at most can cause dehydration. Be sure to also store a manually operated can opener so you can open the cans.

Store any supplies necessary for the care of infants. Mothers under stress, injury, or without sufficient water could lose the ability to nurse an infant. Prepare for this by storing formula, sufficient water, and powdered milk. Other items like diapers, bottles, pacifiers, pre-moistened towels, and any medications should also be stored. Make preparations for baby bedding and blankets, too.

If any occupants of the shelter have special needs you must prepare for those needs. Medications, denture supplies, glasses and contact lenses, canes and walkers, hearing aid batteries, and ointments are a few of the items to keep in mind. Consider who would be in the shelter and what items they will need to be comfortable for several days. It is a good idea to write down the dosage schedule of medications so that medications are administered on time and in the right amounts. Keep any doctor or emergency numbers that could be helpful in your shelter. A clock and calendar will be useful for keeping track of time when administering medication.

A word of caution concerning heat and light in your shelter: do not heat or light your shelter with open flames. Candles, oil lamps, gas lanterns, kerosene heaters, and compressed gas cook stoves are disasters just waiting to happen in a sealed shelter. These methods of heating and lighting consume valuable oxygen, give off carbon monoxide, and are incredible fire hazards. Plan to use flashlights for lighting, and store foods that can be eaten cold if no electricity is available.

I have created a few lists below that may help you put your shelter in order. Some items may not apply in your situation while others will. Remember that if you have decided to split your shelter into separate rooms you may have some redundant items from room to room.

First Aid Items:
　　　Portable first aid kit and booklet
　　　Medical scissors
　　　Antibacterial soap
　　　Tweezers
　　　Eyedropper
　　　Burn ointment
　　　Antibiotic ointment
　　　Activated charcoal
　　　Syrup of Ipecac
　　　Potassium iodide
　　　Sterile gauze
　　　Sterile adhesive bandages
　　　Laxative
　　　Antacid
　　　Anti-diarrhea medication
　　　Aspirin and non-aspirin pain relievers
　　　Saline solution
　　　Petroleum jelly
　　　Antiseptic wipes

    Medical tape
    Thermometer

Personal Hygiene Items:
    Toothbrushes
    Toothpaste
    Toilet paper
    Hand soap
    Feminine hygiene items
    Denture care items

Utility Items:
    Flashlight and extra batteries
    Radio and extra batteries
    Pliers
    Hammer
    Screwdriver
    Crescent wrench
    Water and gas shutoff tool
    Manual can opener
    Gloves
    Sheathed knife
    Scissors
    Plastic sheeting
    Duct tape
    Mess kit for each person with eating utensils
    Portable fire extinguishers
    Weatherproof matches
    Clock
    Calendar
    Garbage bags and twist-ties

Clothing and Bedding:
    At least one complete change of warm clothes for each person in the shelter
    Sleeping bags or blankets
    Hats and gloves
    Heavy-duty socks and shoes
    Long pants and long-sleeved shirts

Coats

Ponchos or raingear

Sanitation Items:

Toilet paper

Garbage bags and twist ties

Plastic bucket with lid

Disinfectant spray

Household chlorine bleach

Infant Items:

Formula

Diapers

Medication

Warm blankets

Pacifiers

Bottles

Powdered milk

Diaper rash ointment

Moist towelettes

Warm clothing and hat

Special Needs Items:

Eyeglasses

Contact lenses and supplies

Canes or walkers

Medications and medication schedules

Hearing aid batteries

Emergency numbers for doctors or caregivers

Decontamination Items:

Flour, fuller's earth, or baby powder

Several gallons of plain household chlorine bleach

Sponges

Liquid or bar soap in quantity

Saline eyewash

Water for use in decontamination

Large bucket

Low stool or short platform on which to stand

Paper towels

Garbage bags and twist ties
Soft brush

Finding and Creating Improvised Shelters

Improvised shelters (sometimes called "in-place" shelters) are structures that offer protection from a particular type of threat when you don't have access to a managed or ready shelter. The types of threat that I will be discussing here are blast, radiation, and chemical/biological. Each of these threats presents a different problem that must be solved by a structure if that structure is to be considered shelter. For example, a sealed room made from plastics, glass, and wood may be perfectly qualified to serve as a biological agent shelter. This same shelter, however, would be useless as a blast shelter. An underground room made of reinforced concrete that has open ventilation may be a very effective blast shelter, but would be inadequate in providing shelter from a chemical agent. The shelter you choose must match the anticipated threat.

IMPROVISED BLAST SHELTERS

Blast shelters are those structures capable of withstanding the initial blast and shrapnel that would emanate from a conventional explosive or possibly a nuclear device. Obviously a nuclear shelter will need to be more structurally sound than a shelter for conventional explosives, but there is enough in common between the two that I will treat them as one.

Many countries have built blast shelters designed to withstand nuclear attack. Most of these shelters were built at the height of the Cold War. It is a good idea to familiarize yourself with any of these locations that are located close to where you live or work. These reinforced, managed shelters should be your first choice if you believe a nuclear or conventional attack is imminent.

Most of us, however, don't live or work near a reinforced nuclear bomb shelter. It will be up to you to know what constitutes a reasonable, if not optimal, shelter. Your blast shelter should be able to protect you from the crushing shock wave of an explosion as well as the associated heat and shrapnel. Materials that can provide this type of protection are steel, concrete, and thick layers of soil. The thicker the material is, the better.

Places above ground that could make good blast shelters are bank vaults, bridge abutments, big steel dumpsters, and most any substantial structure made of concrete and steel. Underground blast shelters generally will provide much more protection. Underground blast shelters could be basements and cellars, subway tunnels, and culverts. Try to close the entrance of your shelter. If you can't close the entrance get as far away from it as possible. Don't ever stand directly in front of the entrance even if it is sealed. Try to place yourself out of the direct line of any blast wave that may enter through any opening in the shelter. Remember to check overhead as well. If you are unable to get out of the direct line of an opening where the blast wave may enter, then move as far away from the opening as possible and face away from the entrance. Squat down and place your arms and hands over the back of your head and neck to protect them.

IMPROVISED RADIOLOGICAL SHELTERS

Improvised radiological shelters are those structures capable of protecting the inhabitants from radioactivity in the environment in the absence of managed shelters. Again, many countries built radiological shelters during the Cold War. If any of these shelters exist near your home or work then they should be your first choice if you have to take shelter.

In the absence of a managed or ready radiological shelter you will need to find sufficient improvised shelter quickly. Radiological agents do not have to come into direct contact with you to be harmful. The harmful effects of radiological agents come from the radiation produced by the agent. The radiation comes in the form of alpha, beta, and gamma rays. Of these, gamma rays are clearly the most dangerous to the human body. Gamma rays can penetrate some materials like glass, plastic, wood, and sheetrock very easily. They have a much harder time penetrating concrete, metal, and soil. As the thickness of the material increases so does its effectiveness as a shield. You should keep this in mind when looking for shelter from a radiological agent. The ability of radiation to penetrate many materials is an important difference between radiological agents and chemical/biological agents.

When seeking shelter from a radiological agent you should look for thick-walled structures made from concrete or steel. If you can find an underground shelter that doesn't have open ventilation to outside contamination you may be in luck. Underground facilities are especially good protection from radioactive contaminants. Barring that, you should move to the innermost section of the structure you choose. Be mindful of the overhead protection as well. Radiation penetrates roofs and ceilings as well as walls. If you take shelter in a multistory building, take refuge low in the building and as close to the center as possible. Shut off any ventilation into your shelter and seal every opening using duct tape and plastic sheeting if it is available.

If you must take shelter in a structure whose walls are exposed to the contamination outside of the shelter you should not stay near the walls themselves. Move toward the center of the room and get as much distance between you and the exposed wall as possible. This will reduce your radiation exposure.

If you are caught in the open when a radiological agent is distributed you should seek shelter in the best place you can find. Take advantage of bridge abutments if available. Move away from the sides of the bridge and place yourself tightly into the abutment.

You may also be lucky enough to find a culvert or some similar underground hollow. Underground cover provides quite good protection since radiation does not penetrate thick soil very effectively. Be careful in culverts and tunnels. There are hazards in them sometimes that can be as deadly as the radiological agent you are seeking shelter from. Snakes, animals, and swarms of insects inhabit many underground lairs. Be mindful, too, that a culvert and any drainage tunnel may quickly become filled with water. Stay alert and be ready to vacate the shelter at a moment's notice. Move as far away from the opening of your underground shelter as you think is safe. Most of the radiological agent will be scattered around the entrance, so this is an area to avoid.

Radiological agents sometimes lose their potency within a few hours or days. The amount of time that a radiological agent is considered dangerous is dependent largely upon the kind of agent it is. It is very helpful to have a portable radio with you. A portable radio could be your sole source of information about what type of agent was used, the length of time you must remain in your shelter, and what route to take to safety. If your radio is unable to receive stations while it is deep in your shelter, you might want to try placing the radio near the entrance. If it works in that position leave the radio there, turn up the volume, and listen to it from deeper within the shelter. The radio may become contaminated since it is so close to the entrance. Make sure you are well away from the entrance while listening, and take care not to become contaminated by the radio if you retrieve it.

IMPROVISED CHEMICAL AND BIOLOGICAL SHELTERS

Improvised chemical and biological shelters are those structures capable of protecting inhabitants from chemical or biological agents in the environment when a managed or ready shelter is not available. One of the most important aspects of a chemical or biological shelter is its ability to keep air in the environment from reaching the inside of the shelter. Given the potency of modern chemical agents, even a miniscule amount of contaminated air could prove deadly to the occupants of a poorly sealed shelter.

Most chemical agents are heavier than air. The troops who fought in World War I found this out in the most disastrous way when chlorine gas crawled along the ground and then fell like a waterfall into the trenches. When searching for an improvised shelter against a chemical attack, you may be well served to find a shelter above ground level by at least several feet.

If there are ventilation outlets in your shelter you should seal them with plastic sheeting and duct tape if you have it. If you don't have duct tape and plastic you should use anything available to cram into the vents to stop the flow of air. If you are able to shut down the ventilation fans that would blow air into or pull air from your shelter then do so.

Seal broken, leaky, and large windows first then move on to more secure and smaller windows. Use anything to block the flow of air in any structural cracks, unused electrical outlets, and the cracks around door facings. Don't use metals (such as aluminum foil) around electrical outlets or wires, but it is fine for cramming into cracks. Anything is better than nothing. Use decontaminated clothing, rags, or paper to cram under the door. Lock the door to your shelter to discourage any person from opening the door without having performed proper decontamination procedures.

Many chemical agents dissipate in a matter of minutes or hours. Some, as discussed earlier, are more persistent. The organisms in biological agents often die within hours or days, and some are killed by long exposure to sunlight. Others, like anthrax, are more persistent. Stay in your shelter until you are informed that it is safe to leave, or you are forced out because of environmental factors.

OTHER IMPROVISED SHELTER CONSIDERATIONS

Don't use open flames to heat or light your shelter. An open flame uses up a lot of valuable oxygen, could set fire to your shelter, and can create noxious fumes. The risk of an open flame in your shelter just isn't worth the price you may have to pay.

Oxygen availability and carbon dioxide contamination can be a concern in shelters that are well sealed. An adult at rest will require about three cubic feet of air per minute. This means that in a perfectly sealed room with an eight-foot ceiling, each person will require just over ten square feet of floor space for five hours. A seven-foot ceiling would allow a little less than four hours of air per person for the same ten feet of floor space. Likewise, two people in the same space cuts the time in half, three people will cut it to one-third. When occupying an improvised shelter the oxygen and carbon dioxide exchange must be considered. If you are in a position in which there is only a small space, and there are several people occupying it, you should be aware of the amount of oxygen available in the room. Keep this in mind when occupying sealed spaces. If only smaller rooms are available it will be more efficient to place a few people in each room.

Hygiene is an issue in any shelter, especially if you will be there for a few days. Human waste must be collected and disposed of in a safe manner. If you have plastic bags available, use them for human waste. Reuse the waste bags if you must, but keep the outside of the waste bag clean. Human waste may contain harmful viral or bacterial agents. Use disinfectant spray on surfaces that become soiled. Dispose of the waste outside your shelter when it is safe to do so. If no bags are available, use trashcans, buckets, plastic containers or any other vessels to collect waste. Cover the containers with anything available when not in use. Being in the presence of bodily waste is not pleasant. You must remember that you are in an emergency situation in an improvised shelter. Do the best you can and adjust your attitude to match the severity of the situation at hand.

An improvised shelter may not have a supply of water. A person can survive for up to five days without water under normal conditions. Environmental temperature and personal activity can have a significant impact on the amount of water you need each day. Even talking can increase the amount of water you need since every word you speak causes you to exhale moisture. Keep your mouth closed to reduce the amount of water your body will require. Breathe through your nose. Conserve any water you may have in your shelter but be careful not to ration drinking water. Don't be wasteful, but drink when you are thirsty. Drink small amounts at regular intervals and do not gulp down water all at once.

If you have entered an improvised shelter that has running water you must remember that the water could be contaminated. Water sources in pipes generally will be safer than water in the open, such as ponds and lakes. Runoff water is the water most likely to have a high concentration of contamination. Any water you intend to drink should be inspected thoroughly. Hints of contamination might include oily spots on the surface or the scent of garlic, almonds, or mustard. Most biological agents will not be detectable. Under no circumstances should you drink water that shows any hint of contamination. If you have access to a water heater that is sitting in a sheltered environment (such as the basement of your house) you can usually assume that the water in it is safe. Drain the water heater into a clean container, cover it and take it to your shelter. I explain exactly how to do this in the next chapter.

Food may not be available in an improvised shelter. A person may survive for up to thirty days without food under most conditions. Water is your first priority, but food provides energy. You may feel lethargic if you don't have food, but it is not immediately life threatening. Consume any safe food you have in small amounts at regular intervals to stave off hunger.

Water Storage, Procurement, and Decontamination

Water is a necessity for living your life under normal circumstances, even when all is well. In an emergency, however, the need for water takes on greater significance where life and death hang in the balance. Because of the need for rapid and effective decontamination, water becomes especially important in terrorist attacks where distributed agents have been released. It is crucial that you know how to find, store, and decontaminate water for use in dire situations.

WATER STORAGE

Water should be stored in portable plastic containers, not glass. Glass breaks too easily and adds unnecessary weight. If any single water container holds more than a couple of gallons, the filled container may be too heavy for some people to transport easily. Keep your container sizes to two gallons or less. Another advantage to storing your water in smaller, separate containers is that if a container is spilled or becomes contaminated, you still have other containers ready for use. Store the water in a cool place that is dark. Each water container should be sealed tightly and the date you sealed it should be written on the container. Empty, clean, and refill the containers every six months or so.

At the very least you should store enough water to last three days for each person in the shelter. As I've said before, a two-week supply of water would be ideal, but may not be practical in some situations. Each person will use about two quarts of water each day under normal conditions. Environmental fluctuations and the activity of each person may vary this amount widely. In an emergency one quart per day per person is sufficient. One gallon per day per person is optimal and is enough for substantial comfort including hygiene if used sparingly. Whether you store a quart, two quarts, or a gallon per day per person will depend upon your particular shelter's capability to store water. In any case, store as much water as practically possible. Remember that the water you store must be enough to meet the requirements for each person who will occupy the shelter. Also, you should store enough to perform decontamination should you become the victim of a distributed agent attack. The more water you have, the better your chances of survival.

WATER PROCUREMENT

Should you find yourself without water there are several places you may be able to procure enough for your survival. These sources may provide water, but the water itself may not be suitable for drinking. If you intend to drink the water you will need to decontaminate it first. I discuss decontamination below. Don't waste time and supplies decontaminating water you do not intend to consume. Murky pond water may work just fine for cleansing your body so long as it does not contain a harmful distributed agent. This same water, though, would not be a good candidate for drinking before being decontaminated.

If you are inside your home, shut off the water supply at the source and use only the water that is stored in the pipes and water heater already in your home. Water that comes from outside your home is suspect and should be considered contaminated until civil authorities tell you otherwise. If the attack was against a water plant or wells in your area, you should not use water in your home at all. It may have already been contaminated.

To extract water from pipes or the hot water heater in your home you should first shut off the main water coming into the house. After this is done go to the highest cold faucet in your home and turn it on. If you have a two-story home, open a cold faucet on the second story. A little water may trickle out, but it won't be much. Then go to the lowest cold faucet in your home and open the faucet. Water should begin flowing and will continue until the pipes are empty. Other faucets may provide more water as well depending on how pipes are routed in your home. To extract water from your hot water heater, you should go to the highest faucet again and open the hot faucet. There will be a small flow from the pipe but it will quickly diminish. Then go to the hot water heater and open the drain on the bottom of the tank. The water could be very hot, so be careful. Capture the water in a container and take it to your shelter.

Another supply of water in your home or office could be in the freezer. Ice cubes and bags of ice can be melted to provide safe drinking water. In a real crunch you can take water from the tank of the toilet. Do not use water from the bowl of the toilet for drinking. Also check for any items such as sodas and juices.

Some canned vegetables like corn, green beans, potatoes, beets, and carrots contain a lot of water. Use canned sources, but be sure to decontaminate or otherwise clean the outside of the cans before opening them so that you don't contaminate the water they contain.

Running water from rivers and streams must always be decontaminated before drinking. If you are in an area that has been subjected to an attack by a distributed agent, the water is almost certainly contaminated. The same is true for water taken from lakes, ponds, runoff water, collected rain, and melted snow and ice. Any of these sources may be used in an emergency, but try to avoid using water that has a scent or floating debris present.

Most of the water on the planet is in the oceans. Water in the ocean is as plentiful as it is salty. Don't drink water straight from the ocean or a salt sea. Use the method described below to remove salt from water. The process is called desalination.

Follow the decontamination procedures described below on all water obtained from sources that are outdoors or otherwise should be suspected of contamination. Heed the warnings, too, that some contaminants cannot be removed or neutralized safely.

SALTWATER DESALINATION

Desalination is the act of removing salt from saltwater. There are various methods of accomplishing this, but the desired result is the same. You want to extract fresh water from salt water.

Desalination filters exist and can be purchased from a variety of manufacturers. Marine supply stores have desalination filtering systems for use on boats. Most desalination filter systems are too bulky to be truly portable in an emergency. While desalination filters are designed specifically to remove salt from seawater, they are not necessarily designed to remove chemical or biological contaminants.

There is, however, a simple and effective way to remove the salt and most contaminants from seawater. The process is called distillation and it also removes other impurities and particles. Distillation involves collecting the water that forms when steam comes into contact with a surface. There are many ways to do this and you may devise effective ways on your own. One method is to boil water in an open container, let the steam deposit itself on a mirror or piece of metal, and catch the water that drips from the metal or mirror in a cup. The piece of metal could even be the lid of the pot you are using to boil the water. The water collected in the cup is salt-free and should be safe to drink or use as an emergency eye-flush.

Distillation is an effective way to decontaminate water because it removes most salts, chemicals, and radiological agents while killing virtually all of the biological agents. Distillation is a simple but slow process. On the other hand, it's also something that you can practice in your kitchen to perfect your technique. You are dealing with boiling water and hot steam, so be careful.

WATER DECONTAMINATION

Water decontamination is the act of making water safe to drink. Decontamination is accomplished by removing the contaminant through filtration or, alternatively, by rendering the contaminant inert while it remains in the water by changing it in some way. It is important to understand that some chemical and radiological agents cannot be removed from water safely by methods described here. The decontamination steps described here are effective against most biological contaminants. Learning to decontaminate water is a valuable skill. It will allow you to take water found in the environment and render it safe for drinking. The desalination method described above is also a decontamination method. The methods below are strictly decontamination procedures and they will not desalinate saltwater.

Water may be boiled to kill most biological agents, but it is not a reliable method of decontamination where chemical and radiological agents are concerned. Bring a container of water to a rolling boil and keep it there for twenty minutes. Some of the water will evaporate. Cool the water before attempting to drink it. The poor taste of water that has been boiled is due to the lack of oxygen. To replenish the oxygen in previously boiled water just seal a half-filled container of the water and shake it vigorously. The water will absorb the oxygen from air in the container and the taste will greatly improve.

Decontamination can be accomplished to some extent chemically. To chemically decontaminate water you can add a few drops of plain household chlorine bleach. Check the label of the household bleach to make sure it contains around 5.25% sodium hypochlorite, and that it is not scented or altered to make the bleach safe for using with colored laundry. Add fifteen to twenty drops of bleach to every gallon of water to be purified. Make sure you stir it or shake it to distribute the chlorine into the water. If you are using a container with a screw top, be sure to leave the screw top a bit loose. It should be loose enough to allow a little water to slosh out around the threads. This will help decontaminate the threads and lid by dousing them with decontaminant. Let the water stand for thirty to forty-five minutes. Smell the water. If the water does not have a hint of chlorine smell, you must repeat the process until it does. The smell of decontaminated water should be similar to that of a chlorinated pool, but not quite as strong. This method of decontamination will do nothing to remove chemical or radiological agents in water, but is effective against many types of biological agents.

Manual filtration is yet another viable way to decontaminate water. Water filtration involves passing water through a material that is perforated with openings large enough to allow the passage of water, but too small to allow particles in the water to pass. The smaller the openings in the filter material, the more particles the material will remove. There are various methods and machines that can accomplish the filtration task. If you decide to purchase a water filter system for decontamination purposes, you should buy a filtration system that uses multiple filters designed to catch particles of ever-decreasing sizes. Single-stage filters clog very quickly and become useless in short order. Bacterial filters are rather common and are quite effective. The physical size of the smallest bacteria, however, is roughly the size of the largest virus. Filters capable of removing viral agents in water must be extremely fine-grained. Reverse osmosis is probably the most effective method of decontaminating water by filtration. High-quality reverse osmosis filtration is capable of removing heavy metals, chemical, biological, and viral contaminants.

A final note on decontamination: the best method to use is the one that is effective against the contaminant. Become familiar with all of the methods described here. Then, when a terrorist attacks, you can gather information from the radio that will allow you to determine the best method to use. Sometimes, just to be safe, you may wish to use more than one method to decontaminate your drinking water. The best water under any circumstance is the water that does not require decontamination in the first place. Take steps to ensure that you have plenty of safe water stored in your shelter.

Choosing and Using Radios and Telephones

In any terrorist attack it is of great importance that you keep informed of all developments. The information provided through radio and telephone can save your life. There are a few things that you should keep in mind when choosing or attempting to use communications equipment. I'll cover some of the issues here.

I have mentioned throughout this book the use of portable radios to gather information. The simple AM and FM commercial radios are quite good for this purpose. I personally stay away from any radio that uses a 9-volt battery because these types of batteries tend to drain quickly. I prefer a radio that uses two or more AA batteries. Don't be fooled into thinking a small 1.5-volt volt battery won't last long. The efficiency of the circuit design determines how long batteries will last, but AA batteries generally pack about ten times as many milliamps as do typical 9-volt batteries. Select a radio that can operate for a very long period of time before needing new batteries.

There are radios that have hand-crank and solar cells that may be used as alternative power. Some of these radios also use batteries as backup. While these are great ideas I think it is wise to take into consideration a couple of facts. Solar powered radios require light to use the solar cell. You may not always have enough light to power the radio. Hand-crank radios are typically much larger and heavier than comparably equipped radios that use batteries alone. This cuts down on portability. Each type of radio has its use, but you must be aware of the weaknesses and strengths of each one and choose accordingly.

Your portable radio should at least receive AM and FM. You might choose a radio that will receive shortwave broadcasts as well. Shortwave receivers are more expensive and can be harder to operate, but they can give you a wide range of alternatives for receiving information. Do not select a radio that has a million reception bands that eats batteries rapidly. It is more important to have a small, simple to operate, efficient radio. Remember that you may have to use your radio in total darkness. Too many buttons can get in the way of operation. Keep it simple.

If you are using a digital radio, make sure you program in your local radio stations most likely to be good sources of information in an emergency. In a terrorist attack all radio stations will probably be announcing informational bulletins, but some may do so more often than others. Make your presets on both AM and FM. Since you could be on the move or in a shelter some of the time you should make choices that are more powerful stations, and not university or low-power special use stations. Choose stations that are geographically widespread so that one terrorist attack does not eliminate every radio station you've programmed.

If you have access to a managed shelter or a room you plan to use as a ready shelter, you should take your radio into the shelter to make sure you can receive the stations you have chosen. You may need to walk around the shelter to find the spot that has the best reception. If you can't receive the stations you've chosen, then try other stations. If you can't receive any stations, you should rig an external antenna to boost your ability to receive stations. Sometimes simply wrapping a long, bare wire around your antenna is sufficient. If you decide to run a wire that extends outside your shelter, you must take care to seal the area where the wire goes through your wall or window. Contact a local amateur radio operator if you have questions about rigging antenna wires. They are quite good at this type of installation and they take great pride in getting good reception from weak signals in poor conditions.

Keep spare batteries for your radio handy. Tape them to the back of the radio if you must. Double check the type of batteries you buy to make sure they are right for your radio and also that you have enough batteries to replace all of them at the same time. Avoid mixing old and new batteries because old batteries will cause new batteries to lose power quickly. In an emergency, however, do what you must to keep the radio working.

When you need to communicate with the outside world from inside your shelter, a telephone should be your first line of communication. You should place a telephone wall jack inside your ready shelter. It is possible for the electrical power to fail while the phone line is still operational. When selecting a phone for your shelter you should select a phone that does not require power from a wall socket to operate. Phones that require power from wall sockets are useless if the power goes out. Portable phones should also be avoided for this same reason. Even though portable phones contain a battery, the base station that receives the signals requires external power. Get the simplest phone that plugs directly into the phone jack. These phones are inexpensive and very reliable.

When communicating while on the move, most people use cellular phones. These devices are great for everyday use and may be good for emergency situations as well. Keep in mind that cellular phones may not work well in shelters below ground or in some buildings. Most cellular phones have a little reception meter that tells you whether the phone is able to receive and transmit signals. Look at the meter and don't waste your batteries if the meter indicates that there is no signal. The telephone companies build the circuits to handle a certain number of calls. In an emergency it is possible that the circuits could become overloaded with calls and you may not be able to get a call through even when you have a signal. Be patient. If you are running low on battery power you should shut off your phone and try again in thirty minutes or an hour.

Besides cellular phones there are other forms of communication available. Family and citizens band radios are two alternatives. Family band radios have a limited range of around two miles but they could be useful in an emergency. Some family band radios even have built in AM or FM receivers. Citizens band radios, also called CBs, have a much longer range under most conditions. Truckers use CBs extensively and there is usually someone listening on nearly every channel.

Another alternative is amateur radio, also called "ham radio". Don't let the name "amateur" fool you. A great deal of radio and antenna knowledge is found among the ranks of amateur radio operators. Amateur radio operators have provided crucial communications in countless emergencies and disasters all over the world. To become an amateur radio operator you must pass a test on your knowledge of radio and antenna design and operation before receiving a license. There are many license levels and the higher you go the more you have to know. The advantage to amateur radio is that to be an operator you must know something about how radios and antennas work. In an emergency this can prove to be a vital skill for you and others around you. Another advantage is that it gives you the privilege of operating radios that are more versatile and much more powerful than unlicensed operators are allowed. Amateur radio operators communicate with operators on the other side of the world as a matter of routine. Becoming an amateur radio operator allows you to provide a valuable service to your community and country.

Skills You Should Acquire

The subjects I briefly discuss below are skills I believe are important and should be learned by every American. I list them here because, though very important, they are beyond the scope of this book. Agencies dedicated to teaching these skills have perfected techniques to help you learn and apply what you've learned in an emergency situation. These skills come in handy not only during a terrorist attack but also in situations that people face every day across America. You will be safer and the people around you will be safer as well.

FIRST AID

In a terrorist attack or in everyday life a basic knowledge of first aid can mean the difference between life and death to you or someone you know. You should seek proper training from the American Red Cross or similar organization. There you will learn the skills to stop bleeding, clear air passages, administer CPR, bandage wounds, and treat burns and shock. The knowledge you gain could prove to be invaluable in an emergency.

BASIC FIREFIGHTING

Terrorist attacks and simple accidents in your home or shelter can cause fires. Most fires start out small. Small fires can become raging infernos with astonishing speed. Learn how to fight fire and use fire extinguishers. Check with your local fire department to find out where basic firefighting training is available in your area.

BASIC SURVIVAL

Any knowledge you obtain about surviving and orienting yourself in the wilderness could be very valuable in any situation where you are left without the comfort and familiarity of home. Even if you don't find yourself in a wilderness setting, the knowledge you gain about procuring food, water, shelter, and cooking without access to a kitchen can increase your comfort at least, and at most it could save your life. A basic knowledge of rope and knots can be quite useful as well. Learning the techniques of basic survival can be interesting and entertaining for everyone involved.

SWIMMING

No matter where you live, swimming is a skill that can come in handy. Even in the desert, flash floods kill many people. Given the fact that two-thirds of the planet is covered by water it seems reasonable to learn to swim. If you become trapped in a situation where the only escape route is through water, you will be much more likely to survive if you know how to swim. It is never too late to learn to swim, and it's a great, low-impact exercise as well.

SELF DEFENSE

I hold a black belt in Tae Kwon Do. I obviously am not at all opposed to the disciplines of martial art. On the other hand, I am a real skeptic of self-defense classes. I have seen many classes that advertise in such a way that you would believe that in one afternoon (at the cost of $100.00) you can learn enough hand and foot techniques to do real damage to a would-be attacker. Worse yet, these classes are usually full of women who weigh in the neighborhood of 100 pounds and have never struck a blow against another human in their lives. Some of these women, having no prior background in hand-to-hand combat, walk away with a very false sense of security. I can't count the times I've heard women straight out of a self-defense class say they are no longer afraid to walk across that deserted parking lot at night alone. My thought: not so fast.

If the instructor told them that they no longer should be wary and may now boldly go into darkened lots with confidence, then he simply gave bad advice. If he gave such poor advice on such a simple concept, how much worse was his instruction in the intricate maneuvers of hand techniques? The fact is, that even with all of the foot stomping and groin kicking she has learned, she might still be a 100 pound woman without a street fight under her belt going up against a 220 pound hardened criminal who has been in six bar fights this week. She's going to get hurt, and maybe killed, because of her newfound false confidence. My point is this: if you have only 4 hours to spend learning something that can save your life, spend it learning what to look for and what to avoid to stay safe.

Now, having said all that, if you care to invest a little more time, like 4 hours per week for six months, it very well may be worth it. Better yet, commit yourself to making a self-defense oriented class a part of your physical fitness training. Thai kickboxing, Tai Kwon Do, and Ju-jitsu are three art forms that come to mind when thinking of defense and physical fitness. There are many, many other martial arts forms as well, and one of these not mentioned may suit you better. The martial art you choose is less important than the quality of training you receive. Only spend your money and time in reputable instruction from an expert instructor.

In the end, the advantage these forms of self defense offer is that they take time to create mental discipline and confidence, hand-to-eye coordination, and techniques that can't be learned in 4 hours or even 4 weeks. A couple of years of real training three times a week are needed before you can confidently execute hand and foot techniques with effective speed and accuracy in the heat of the moment. The gym is one thing, the street is quite another, and facing an armed terrorist is a whole different scenario.

Conclusion

Throughout this book I have discussed various ways and means by which terrorists may attempt to harm Americans, and some actions you can take to defeat them. Obviously, many other nationalities and ethnic groups are also targeted by terrorists. Israelis and Russians come to mind. Certainly the lessons shared in this book can serve others who find themselves the targets of terrorists, regardless of their nationality or ethnicity. But mine is an American perspective, and it is from this perspective that I have written this book.

We, as Americans, have been the targets of terror for decades. Americans are big-game trophies to terrorists on the ever-widening terrorism battlefield. Terrorism against Americans is not a recent phenomenon, and it is not the result of any covert action or outright wars fought in recent years. Though we have been the targets, the acts of terror occurred mostly abroad. Until September 11, 2001, our soil had largely been left unturned by international terrorism. As horrible as it was, and as well as our war against terrorism has gone, the worst may be yet to come. We have not seen suicide bombers on our buses or in our malls. We have yet to experience car bombs exploding with frequency on our crowded downtown streets. We have not yet had massive releases of deadly chemical or biological agents in our subways and cities. We have, by all accounts, been very lucky.

The day may come when we will not be so fortunate. We need not think that the terrorists will be stopped at the border. Our borders are long and porous. If we can't stop the flood of illegal farm workers from crossing our southern border, we have no reason to believe that we can stop a handful of highly motivated, well-trained terrorists from crossing that same border. If we can't stop the massive flow of thousands of tons of drugs from pouring across our borders and onto our streets, what chance do we have of stopping a few pounds of deadly biological agent from entering our country in the same manner? You do the math.

The threat of terrorism doesn't start at our borders and, as I have just described, it doesn't stop there either. Terrorism is a worldwide problem that requires the United States to respond with political, economic, diplomatic, and military measures. This is a difficult balance, but one that must be struck. Our only alternative is the erosion of personal liberty and freedom at home, and the risk of appearing impotent in the face of terrorism abroad. As Americans, we must insist that our government do whatever is necessary to bring this scourge against us into focus and to destroy it wherever it is found.

Here at home, Americans may soon be on the front lines of terrorism again. This is a war that will be waged over a period of years and, in fact, may never be completely over. Knowing this, it is imperative that Americans adjust to a state of alertness to which we have not been accustomed in the past. Pay attention to your environment, stay informed about current events, keep abreast of information distributed by our government and, above all, use your brain. Every terrorist on our soil is an enemy, and every American has a responsibility of defeating him wherever he is found.

To my fellow Americans, I encourage you to stay strong, stay alert, stay smart, and stay the course of the world's greatest nation.

Index

environment 3, 25, 30, 31, 32, 33, 34, 35,
  36, 52, 64, 72, 73, 74, 76, 77, 78, 79, 80,
  81, 82, 83, 84, 85, 86, 89, 92, 93, 101,
  103, 105, 109, 118
event of interest 33, 34, 36
explosives attacks 22, 55, 56, 60, 62, 64

F

fallout 59, 60, 73, 74, 75, 76
family band 113
filters 108, 110
filtration 89, 94, 109, 110
firearms 49, 50
flour 86, 98
food 25, 71, 76, 82, 83, 93, 94, 95, 105,
  115
fuller's earth 86, 98

G

gas mask 72, 80
gasses 45, 46, 76, 80, 86, 87
grenade 55, 56, 57, 58

H

ham radio 113
handguns 49, 50, 51
harassing agents 77
hard targets 23, 24
hemorrhagic fevers 81
hostages 5, 6, 7, 8, 40, 41
hydrogen cyanide 79
hypothermia 69

I

improvised shelters 80, 89, 100
improvised weapons 52, 53, 54, 70
infants 93, 95
infrastructure attacks 21, 37
instinct 30, 31
item of interest 33

K

knives 16, 27, 46, 47

L

lewisite 78

M

managed shelters 89, 100, 101
Marburg 81
mass transit 21, 25, 35, 37, 55, 63, 65, 68,
  81

N

nerve agents 77, 79
nitrogen mustard 78
nuclear detonation 58, 59
nuclear explosions 58, 73

P

PAL 35, 36
pepper sprays 45, 46
person of interest 31, 34
Personal Alert Level 35
pets 94
phone 17, 61, 65, 92, 112, 113
phosgene 78, 86
phosgene oxime 78, 86
plastic sheets 75, 76, 90
pneumonic plague 82
pointed instruments 53
pralidoxime chloride 80
profiling 1, 31, 32, 33

R

radiological agents 22, 73, 74, 85, 86, 101,
  102, 109, 110
radiological shelters 101
radios 111, 113
rail 64, 65
ready shelters 80, 89, 90, 91, 94

0-595-74912-7